Optoelectronics Circuits Manual

Dedication

To my Lady Esther, the light of my life, who brightens my days and illuminates my nights, with special thanks for the kindness and patience that she has shown throughout the ten months that it has taken me to write this book and to design all of the circuits and generate all of the artwork that it contains.

Optoelectronics Circuits Manual

Second edition

R. M. MARSTON

LEARNING
RESOURCES
CENTRE

Newnes

OXFORD AUCKLAND BOSTON JOHANNESBURG MELBOURNE NEW DELHI

Newnes
An imprint of Butterworth-Heinemann
Linacre House, Jordan Hill, Oxford OX2 8DP
225 Wildwood Avenue, Woburn, MA 01801-2041
A division of Reed Educational and Professional Publishing Ltd

 A member of the Reed Elsevier plc group

First published 1988
Reprinted 1990, 1991, 1993, 1996, 1997, 1998
Second edition 1999
Transferred to digital printing 2004
© R. M. Marston 1999

British Library Cataloguing in Publication Data
A catalogue record for this book is available from the British Library

Library of Congress Cataloguing in Publication Data
A catalogue record for this book is available from the Library of Congress

ISBN 0 7506 4166 5

Composition by Scribe Design, Gillingham, Kent

Contents

Preface

Optoelectronics is the fast-growing branch of electronics concerned with the practical application of modern optical (opto) devices such as light-emitting diodes (LEDs), optosensors, fibre optic cables, and lasers, etc. The first (1988) edition of this book presented a host of practical optoelectronics circuits but ignored fibre optic cables and lasers; the present fully revised and greatly expanded second edition of the book remedies this deficiency by devoting two large chapters to these and associated subjects.

This new 60 000-word edition of the manual explains – with the aid of 307 new illustrations – the operating principles of the most important types of modern optoelectronic devices and systems and – where applicable – provides the reader with a wide range of practical application circuits. The manual is divided into twelve chapters. The opening chapter gives a concise description of optoelectronic basic principles and devices. Chapters 2 to 5 show practical applications of LEDs in various types of display circuits, and Chapters 6 to 8 deal with light-sensitive circuits, optocoupler circuits, and brightness-control circuits respectively.

Chapters 9 and 10 are new and important; Chapter 9 takes an in-depth look at the basic nature of light and at the operating principles of mirrors, prisms and lenses, and is an essential lead-in to Chapter 10, which describes the operating principles and practical applications of fibre optic cables and lasers. The final two chapters of the book deal with light-beam alarms and IR remote-control systems.

This book, like all others in the Newnes *Circuits Manual* and *Users' Handbook* series, is specifically aimed at practical design engineers, technicians and experimenters, but will also be of great interest to many competent hobbyists and students of electronics. It deals with its subject in an easy-to-read, down-to-earth, mainly non-mathematical but very comprehensive manner. Each chapter explains the basic principles of its subject and – when appropriate – presents the reader with a wide selection of practical application circuits, all of which have been designed and fully evaluated by the author.

Throughout the volume, great emphasis is placed on practical 'user' information and circuitry; most of the ICs and other devices used in the practical circuits are modestly priced and readily available types, with universally recognized type numbers. All of the book's diagrams have been generated by the author, using a basic 'Corel DRAW 3' graphics package.

In this book, the values of resistors and capacitors, etc., are notated in the International style that is now used throughout most of the western world, but which may not be familiar to some 'hobbyist' readers in the USA. Such readers should thus note the following points regarding the use of the International notation style *in circuit diagrams*:

(1) In resistance notation, the symbol R represents *units* of resistance, k represents *thousands* of units, and M represents *millions* of units. Thus, $10R = 10\Omega$, $47k = 47k\Omega$, $47M = 47M\Omega$.

(2) In capacitance notation, the symbols μ, n (= 1000pF), and p are used as basic multiplier units. Thus, $47\mu = 47\mu F$, $10n = 0.01\mu F$, and $47p = 47pF$.

(3) In the international notation system, decimal points are not used in notations and are replaced by the multiplier symbol (such as V, k, n, μ, etc.) applicable to the individual component value. Thus, $4V7 = 4.7V$, $1R5 = 1.5\Omega$, $4k7 = 4.7k\Omega$, $4n7 = 4.7nF$, and $1n0 = 1.0nF = 0.001\mu F$.

R. M. Marston
Fuengirola,
Spain
1999

1

Basic principles

The word 'optoelectronics' is a recently devised one. It first came into general use during the 1970s as a catchy name that roughly describes the branch of electronics concerned with the practical application of modern optical (opto) devices. In this context, an 'opto' device can be broadly described as one that is designed to function somewhere within the visible or near-visible 'light' sections of the electromagnetic spectrum; in more scientific terms, it is one that operates within the approximate 10nm to 100μm (10 nanometre to 100 micrometre) wavelength section of the normal electromagnetic spectrum.

Figure 1.1 shows details of the electromagnetic spectrum, and *Figure 1.2* shows the so-called 'visible light' part of the spectrum, with all wavelengths marked in decade multiples or submultiples of the metre. Note in *Figure 1.2* that the precise upper and lower limits of the visible light part of the spectrum vary between individuals, but typically span 400nm to 700nm; a white light is produced when all the colours of the visible spectrum are present simultaneously; a white light can easily be filtered to reproduce any desired individual colour.

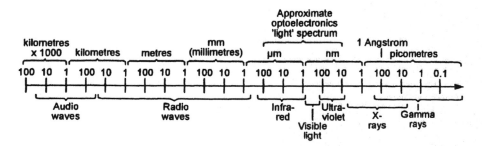

Figure 1.1 *The electromagnetic spectrum (wavelengths given in decade multiples and submultiples of the metre)*

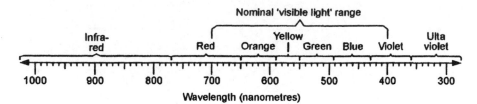

Figure 1.2   *The visible-light part of the spectrum (wavelengths in nanometres)*

The area of interest in optoelectronics spans the entire visible light part of the *Figure 1.1* spectrum, plus most of the infra-red (IR) and ultra-violet (UV) ranges. All signals within this span have characteristics like those of normal visible light; they pass easily through optically clear mediums, can be completely blocked by optically opaque objects, and can be manipulated by optical lenses. Consequently, in optoelectronics, IR and UV radiations are often referred to as 'invisible' light.

Optoelectronic devices and circuits have many practical applications. They can be used to generate a wide variety of visual information displays, or to give an automatic switching or alarm action in the presence or absence of a visible or invisible light source, or to give a similar action when a person or animal moves within the range of an IR 'body heat' detector. They can be used to give remote-control action via an infra-red light-beam, or may use a fibre optic cable as a low-loss interference-free data link between one circuit and another.

The 'opto' devices used in optoelectronics come in four basic types, which can be described as optogenerators, optosensors, optocouplers, and opto-reflectors. Each of these types of device may, in turn, come in a variety of basic forms. The remaining parts of this chapter describe the most widely used types of optical devices, and basic ways of using them.

Optogenerators

Introduction

An optogenerator is a device that, when suitably stimulated, generates visible or IR or UV light. The most widely used types of man-made optogenerators are the ordinary filament lamp or light bulb, the light-emitting diode (LED), neon lamps, fluorescent displays, and lasers. This section presents basic details of these types of optogenerator device.

Note than many natural objects (including the sun) inherently act as optogenerators. IR radiation is, for example, a natural by-product of heat radiation, and all warm-blooded creatures (including humans) thus act as

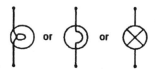 or or

Figure 1.3 *Alternative circuit symbols used to represent a filament lamp*

natural IR generators. The modern security industry takes advantage of this fact by using sensitive optoelectronic passive infra-red (PIR) sensors to monitor an area and activate an alarm if the sensor detects the presence of a human intruder via his/her body-heat IR radiations.

Filament lamps

The ordinary filament lamp or light bulb is the most widely used type of optogenerator device. It can be powered from either AC or DC voltage sources, is usually used as a 'white light' source, has a multitude of practical applications in the home, in automobiles, and in industry, and can be represented by any of the basic circuit symbols shown in *Figure 1.3*. It usually comprises a coil of tungsten wire (the filament) suspended within a vacuum-filled glass envelope and connected to the outside world via a pair of metal terminals; the filament runs white hot when connected to a suitable external voltage, thus generating a bright white light.

The filament lamp has three notable characteristics. One of these is that the filament's resistance has a positive temperature coefficient, causing the lamp resistance to increase with operating temperature; *Figure 1.4* shows the typical resistance variations that occur in a 12V 12W lamp. Thus, the resistance is 12 ohms when the filament is operating at its normal 'white' heat, but only 1 ohm when it is cold (i.e. at normal room temperature). This 12:1 resistance variation is typical of all tungsten filament lamps and causes them to have switch-on 'inrush' current values about twelve times greater than their normal 'running' values.

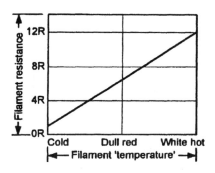

Figure 1.4 *Typical resistance variations that occur in a 12V 12W filament lamp*

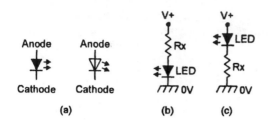

Figure 1.5 *Alternative circuit symbols used to represent a LED (a), and alternative ways of limiting a LED's operating current via a series resistor (b) and (c)*

The second notable feature of the filament lamp is that it has a fairly long thermal time constant, and power thus has to be applied to (or removed from) the filament for a significant time (tens or hundreds of milliseconds) before it has an appreciable effect on light output. This characteristic enables the device to be powered from either AC or DC sources, and enables the lamp brightness to be varied by using highly efficient switched-mode pulsing techniques.

The final notable feature of the filament lamp is its very poor electrical-to-light energy conversion efficiency, which is a mere fraction of one percent, and even then three-quarters of this minute radiated output is in the invisible IR range and only the remaining one quarter is in the visible light range. The human eye typically responds to light wavelengths within the 400 to 700 nanometre range, with its sensitivity peaking at about 560 nanometres, but tungsten lamps give light outputs that typically peak between 800 and 1100 nanometres, in the invisible IR range.

Light-emitting diodes

The most widely used modern *electronic* optogenerator device is the light-emitting diode, or LED. *Figure 1.5(a)* shows standard circuit symbols that represent this solid-state device, which has electrical characteristics similar to those of a normal diode (i.e. it passes current easily in one direction but blocks it in the other), but emits narrow-band light when forward biased. Standard LEDs emit a red light, but others are available that emit orange, yellow, green, blue, or IR types of light.

In use, the operating current of a LED must be limited to a safe value; this can be achieved via a series resistor connected to either the anode or the cathode, as shown in *Figures 1.5(b)* and *1.5(c)*. LEDs have very fast optoelectronic response times, and can easily be used to transmit a variety of coded remote-control light signals, etc.

LEDs generate forward operating voltages of about 2V and typically use drive currents in the range 2 to 30mA. They are available in single-LED packages and in various multi-LED styles. 2-LED packages housing a pair of red and green LEDs are, for example, available in either bi-colour or tri-colour forms; in the bi-colour device only one LED can be illuminated at a

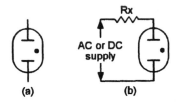

Figure 1.6 *Symbol (a) and basic usage circuit (b) of the neon lamp*

(a)   (b)

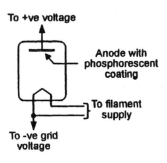

Figure 1.7 *Basic construction of a fluorescent or phosphorescent display device*

time, so the device emits either a red or green colour, but in the tri-colour device both LEDs can be illuminated at the same time, generating a yellow colour in addition to red and green. Multi-LED packages are also available in bar-graph, dot-matrix, and 7-segment alphanumeric display forms (described later in this chapter).

Neon lamps

Neon gas discharge lamps are simple optogenerators that comprise an inert neon gas and a pair of electrodes, housed in a glass envelope. When a suitably high 'striking' voltage is applied to the electrodes the gas becomes conductive, producing a red glow near the electrodes; if the voltage is further increased the glow spreads through the gas. In use, a resistor is wired in series with the neon lamp, so that the neon voltage self-limits to slightly above the striking value. *Figure 1.6* shows the symbol and basic usage circuit of the neon lamp, which can be powered from either an AC or DC voltage.

Fluorescent or phosphorescent displays

Another type of light-generating device is the fluorescent or phosphorescent display, which is shown in basic form in *Figure 1.7*. Here, an incandescent filament acts as a source of free electrons, which can be accelerated into a phosphor-coated anode via a suitable grid-to-anode voltage, thus generating a visible (usually green, blue, or ultra-violet) fluorescent glow.

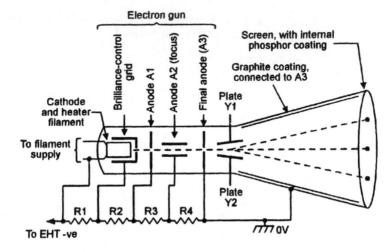

Figure 1.8   *Basic structure of a cathode ray tube (oscilloscope type)*

The cathode ray tube

One of the most important applications of the above-mentioned phosphorescent display principle is in the cathode ray tube (CRT), which is widely used as a TV screen, as a computer monitor screen, and as an oscilloscope display unit. *Figure 1.8* illustrates the basic structure of the oscilloscope version of such a unit.

In *Figure 1.8*, the cathode is heated by a filament and acts as a source of free electrons, which have their flow-density controlled by a grid voltage, and which are accelerated along the tube and compressed into a narrow central beam (the cathode ray) via a series of anodes that are positively charged relative to the cathode; this cathode–grid–anode assembly is known as an *electron gun*. The narrow cathode ray beam that emerges from the electron gun speeds on to the end of the tube (the screen), which is internally coated with phosphor powder and thus produces a pin-point glow of light where it is struck by the beam.

In *Figure 1.8* it can be seen that the cathode ray electron beam can be shifted up and down (on the Y axis) by applying a suitable voltage to a pair of internal 'Y' deflection plates. In practice, the tube contains two pairs of such plates, as shown in the frontal view of *Figure 1.9*; the X plates enable the beam to be moved to the left or right, and the Y plates enable it to be moved up or down. Thus, the bright spot can be moved to any part of the screen via the X and Y plates, which (by using rapid movements) can thus be used to draw an infinite variety of patterns, shapes or numerals, etc., on the screen.

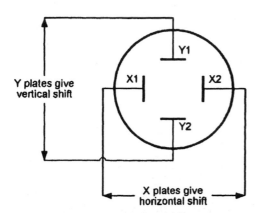

Figure 1.9 *Layout of X and Y deflection plates*

The beam deflection of the oscilloscope-type cathode ray tube described above is caused electrostatically, by the voltages applied to the X and Y plates. A similar deflection can be obtained electromagnetically, be feeding suitable currents to sets of coils mounted on the outside of the tube; this latter technique is used with most TV and monitor tubes. The cathode ray beam can also be controlled on the Z or intensity (brightness) axis by feeding suitable control voltages to the tube's grid terminal.

The laser

Most conventional optogenerators give an output that occupies a fairly wide spread of wavelengths. Such an output cannot be *precisely* focused by a simple lens system, which – at any given focal length – can precisely focus only a *single specific wavelength*; the lens scatters signals of all other wavelengths, producing the effect known as chromatic aberration.

A laser is an optogenerator device that produces a light output at a single (monochromatic) wavelength, which can be precisely focused by a very simple lens system. The output of an infra-red laser that generates an output power of a mere 0.5mW can, for example, be focused into a minute spot measuring only 1.6μm in width (roughly equal to 1/30th of the width of a human hair), in which the focused IR power density has a value (allowing for optical losses) of about 12kW/cm².

The word LASER is an acronym for *L*ight *A*mplification by *S*timulated *E*mission of *R*adiation. The first practical laser was a helium–neon gas type, developed in the late 1960s, that generated a continuous IR beam. Today, lasers are available in many basic types, including gas tube and ruby crystal ones, but the most widely used ones are solid-state 'diode' types, which are represented by the same circuit symbol as an ordinary LED.

Lasers are widely used in modern everyday life. All CD players are fitted with low-power (typically 0.5mW output) laser units that are used to read off digital data from the CD when it is played. Other lasers, with power

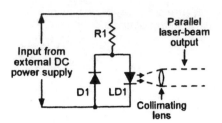

Figure 1.10 *The basic elements of a diode laser unit that gives continuous-wave operation and is powered from an external DC supply*

outputs up to about 5mW, are widely used in bar-code readers, in surveying instruments, in electronic security systems, in special surgical instruments, and in laser pointers, etc.

Complete laser units, consisting of a laser-plus-lens system and typically giving power outputs up to a maximum of about 5mW, are readily available from specialist suppliers. Some units give a visible red output, and others give an invisible IR output; some give only continuous-wave operation, and others have outputs that can be externally modulated.

Figure 1.10 shows – in simplified form – the basic elements of a diode-type laser unit that gives continuous-wave operation and is powered from an external DC supply. Here, the laser diode (LD1) is fed with a reasonably stable drive current via R1, and is protected against damage from reverse-connected supplies via diode D1. The laser diode's optical output is applied to a collimating lens, which convert it into a narrow (typically 1mm to 5mm diameter) parallel beam that can (under low opacity conditions) be radiated for considerable distances with very little loss in power density.

Additional information on laser principles and applications is given in Chapter 10 of this volume.

Optosensors

Introduction

An optosensor is a device that, when stimulated by visible or IR or UV light, produces some kind of electrical reaction. The most widely used types of man-made optosensors are the LDR (light-dependent resistor), the photo-diode, the phototransistor, and the solar cell; *Figure 1.11* shows the circuit symbols of these four devices. Another widely used but very specialized type of optosensor is the pyroelectric IR detector, which forms the basis of all modern PIR movement-detector security systems. This section presents basic details of all of these optosensing devices.

Light-dependent resistors

The LDR is also known as a cadmium–sulphide (CdS) photocell; it is a passive photoelectric device that changes its electrical resistance in the

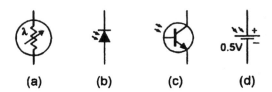

Figure 1.11 *Circuit symbols of (a) LDR, (b) photodiode, (c), phototransistor, and (d) solar cell*

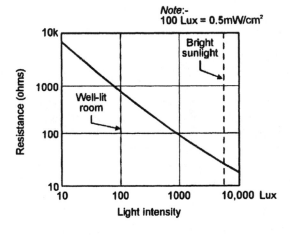

Figure 1.12 *Typical photoresistive graph of a LDR with a 10mm diameter face*

presence of visible light; it has a fairly slow (typically several milliseconds) response time. *Figure 1.12* shows the typical photoresistive graph that applies to an LDR with a face diameter of about 10mm.

The photodiode

All reverse biased silicon semiconductor junctions are inherently photo-sensitive, and are usually shrouded in opaque material to minimize this (normally) unwanted effect. A photodiode is a slightly modified silicon diode that is either mounted in a translucent case or has its semiconductor junction mounted beneath an optical lens, thus exposing it to light. Some photodiodes are made to respond to visible light, and some to IR light.

Figure 1.13 shows a basic way of using a photodiode, by reverse-wiring it in series with load resistor R1. The diode's impedance varies with the illumination level; under dark conditions, the impedance is very high, and negligible current thus flows through R1; under bright conditions, the impedance is fairly low, and hundreds of nA may flow through R1, which thus generates an output voltage proportional to the illumination level.

The phototransistor

A phototransistor works in the same basic way as a photodiode; it is simply a modified silicon transistor with one of its junctions made photovisible.

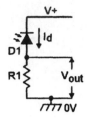

Figure 1.13 *One basic way of using a photo-diode*

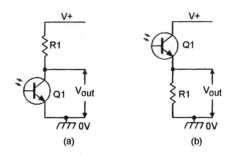

(a)   (b)

Figure 1.14 *Alternative ways of using a npn phototransistor*

Figure 1.14 shows two basic ways of using an *npn* phototransistor. Here, the base–collector junction gives a photodiode type of action, but its photo-sensitive reverse currents are fed into the transistor's base and – by normal transistor action – make a greatly-amplified current flow between the emitter and collector. Load resistor R1 converts this current into a proportional output voltage. A phototransistor's sensitivity is typically about one hundred times greater than that of a photodiode, but its frequency response is propor-tionately lower than that of a photodiode.

Solar cells

So-called solar cells are photovoltaic units that convert light directly into electrical energy. An individual solar cell generates an open-circuit voltage of about 500mV (depending on light intensity) when active. Individual cells can be connected in series to increase the available terminal voltage, or in parallel to increase available output current; banks of cells manufactured ready-wired in either of these ways are known as solar panels. *Figure 1.15* shows how a bank of 18 cells can be used to autocharge a 6V ni–cad battery via a germanium diode.

The available output current of a solar cell depends on the light intensity, on cell efficiency (typically 6 to 12 percent), and on the size of the active area of the cell face. On a bright and sunny day the available light energy, at sea level, is typically in the range 0.5 to 4kW/m², so there is plenty of 'clean' energy waiting to be converted.

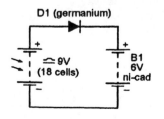

Figure 1.15 *Solar panel used to auto-charge a 6V ni–cad battery*

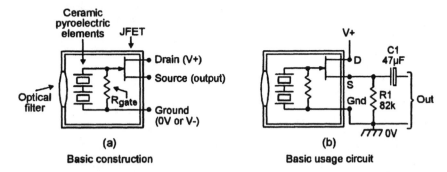

Figure 1.16 *Basic construction (a) and usage circuit (b) of a pyroelectric infra-red detector*

Pyroelectric IR detectors

Some special crystals and ceramics generate electric charges when subjected to thermal variations or uneven heating; this is known as a pyroelectric effect. Pyroelectric infra-red detectors incorporate one or two elements of this type, plus a simple filtering lens and a field-effect transistor (FET), configured in the basic way shown in *Figure 1.16(a)*.

The basic action of the device is such that – if a human body moves within the visual field of its pyroelectric elements – part of the body's radiated IR energy falls on the surface of the elements and is converted into a minute variation in surface temperature and a corresponding variation in the element's output voltage. When the unit is wired as shown in the *Figure 1.16(b)* basic usage circuit, this movement-inspired voltage variation is made externally available via the buffering JFET and capacitor C1 and can, when suitably amplified and filtered, thus be used to activate an alarm when a human body movement is detected.

Note that pyroelectric IR detector circuits of the basic type just described have, because of the small size of the detector's light-gathering lens, maximum useful detection ranges of just over one metre, but that this range can be extended to more than ten metres with the aid of a relatively large external light-gathering/focusing lens of the type used in modern passive infra-red (PIR) movement detector security systems.

Optocouplers

Introduction

An optocoupler is a device or system that can be used to couple the output of an optogenerator to the input of an optosensor. Any practical optocoupler

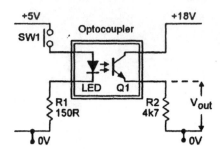

Figure 1.17 *This optocoupler is a closed non-interruptible type*

can be classified as either a 'closed' or 'open' type, with a coupling that either is or is not *physically* 'interruptible'. Optocouplers can thus be designed as either 'closed non-interruptible', 'closed interruptible', 'open interruptible', or 'open non-interruptible' types. This section describes optocouplers of these four basic types.

In the above context, a 'closed' optocoupler is one in which the optogenerator and optosensor are both mounted in a common, closed, housing; an 'open' optocoupler is one in which the optogenerator and optosensor are separately housed.

Closed non-interruptible optocouplers

The simplest optocoupler is the closed non-interruptible type, which typically consists of an IR LED and matching phototransistor, mounted close together and optically coupled within a light-excluding package, as shown in the basic circuit of *Figure 1.17*. Here, SW1 is normally open, so zero current flows through the LED; Q1 is thus in darkness and also passes zero current, so zero output appears across R2. When SW1 is closed, however, current flows through the LED via R1, thus illuminating Q1 and causing it to generate an R2 output voltage.

The R2 output voltage can thus be controlled via the R1 input current, even though R1 and R2 are fully isolated electrically. In practice, the device can be used to optocouple either digital or analogue signals, and in some cases can provide many hundreds or thousands of volts of isolation between the input and output circuits; in the latter case, the device is often known as an 'opto-isolator'.

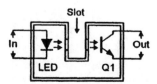

Figure 1.18 *This slotted optocoupler is a closed interruptible type*

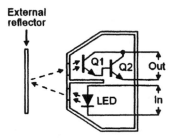

External reflector

Figure 1.19 *This reflective optocoupler is a closed interruptible type*

Closed interruptible optocouplers

Closed interruptible optocouplers are variants of the basic closed type described above, but are arranged so that their opto-link passes out of the unit's housing and then back again via an externally interruptible path. *Figures 1.18* and *1.19* show units of this basic type.

Figure 1.18 shows the basic construction of a slotted or vane-type optocoupler, which has a slot moulded into the package between the LED and Q1. This slot houses transparent windows, so that the LED light can normally freely reach the face of Q1, but can be interrupted or blocked via an opaque object placed within the slot. This slotted optocoupler can thus be used as an object or liquid-level detector.

Figure 1.19 shows the basic construction of a 'reflective' type of optocoupler. Here, the photoactive faces of the LED and Q1 both point outwards (via transparent windows) towards an imaginary point that is roughly 5mm beyond each window, so that the LED light can only reach Q1's face via a reflective surface that is placed at or near this point. This device can thus be used as an external-object detector.

Open non-interruptible optocouplers

The only practical open non-interruptible optocoupler system takes the basic form shown in *Figure 1.20,* in which the output of the optogenerator LED is coupled to the input of optosensor Q1 via a length of fibre optic cable, and

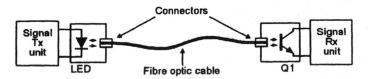

Figure 1.20 *This open non-interruptible optocoupler system uses a fibre optic cable as its 'coupling' medium*

the optocoupling can thus only be physically interrupted by actually sever-
ing the fibre optic cable.

A fibre optic cable can, in very simple terms, be regarded as a flexible light
pipe that can efficiently carry modulated or unmodulated light signals from
one point to another (even if the journey involves bends and loops) with
complete immunity from electromagnetic interference. In most practical
applications the cable must be linked to the light source and destination
points via special connectors, to minimize signal losses.

Two distinct types of fibre optic cable are in common use. One of these is
made from plastic polymer; it can easily be cleanly cut with a sharp blade, is
best suited to use with visible red (rather than IR) light, and is best suited
to short-distance (up to 10 metres) applications; it gives a typical attenuation
of about 200dB/km (0.2dB/metre) when used at its optimum light
wavelength, but up to 1500dB/km (1.5dB/metre) if used with IR wavelength
signals.

The second (and most widely used) type of fibre optic cable uses a glass
fibre construction; it can be correctly cut only with a special (and very expensive) tool, is best suited to use with IR light and to long-distance (hundreds
of metres) use; it gives a typical attenuation of only 1.3dB/km when used
with optimum wavelength IR signals.

Additional information on fibre optic cable principles and applications is
given in Chapter 10 of this volume.

Open interruptible optocouplers

One of the most important applications of infra-red LED/phototransistor
combinations is in open interruptible optocoupler or 'light-beam' systems
such as IR light-beam alarms and remote-control systems. Some elementary
principles of these systems are illustrated in *Figures 1.21* to *1.23*.

Figure 1.21 illustrates the basic operating principles of a simple 'single-
beam' IR light-beam alarm system. Here, the transmitter (Tx) feeds a coded
signal (usually a pulsed squarewave) into an IR LED, which has its output
focused into a fairly narrow beam that is aimed at a matching IR phototransistor mounted on the remotely placed receiver (Rx). The circuit action
is such that, when the light-beam is operating, the receiver's output is

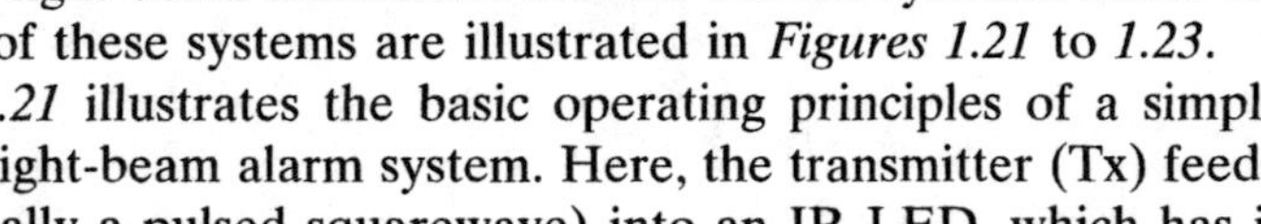

Figure 1.21   *Simple IR light-beam alarm*

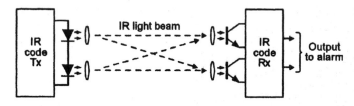

Figure 1.22 *'Dual-beam' IR light-beam alarm*

normally off, but turns on and activates an alarm or safety device if the beam is interrupted by a person or object. This type of system may have an effective detection range of up to 30 metres.

It is important to note that the above system works on a pure 'line-of-sight' principle, in which the actual opto-link can be broken by any opaque object that is bigger than the larger of the system's two lenses and which enters the line-of-sight anywhere between the Tx and Rx lenses (you can demonstrate this effect by looking at a small remote object with one eye closed, and noting how the object can be obscured by simply moving a pencil into the line-of-sight). Thus, this system can easily be false-triggered by a fly or moth (etc.) entering the beam or landing on one of the lenses. The dual-light-beam system of *Figure 1.22* does not suffer from this defect.

The *Figure 1.22* system is basically similar to that just described, but transmits the IR beam via a pair of series-connected IR LEDs that are normally spaced about 75mm apart, and receives the beam via a pair of parallel-connected phototransistors that are also spaced 75mm apart. Thus, each phototransistor can detect the beam from either IR LED, and the receiver's output will thus only activate if *both* beams are broken simultaneously; this will normally only occur if a large (greater than 75mm) object is placed within the composite beam. This dual-beam system is thus virtually immune to false triggering by moths, etc.

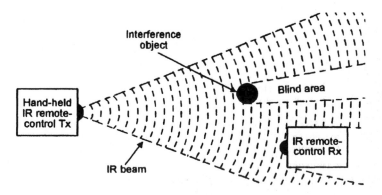

Figure 1.23 *Basic IR remote-control system*

Figure 1.23 illustrates some important operating principles of an IR remote-control system. Here, the hand-held unit transmits a broad beam of coded infra-red light, and can remote-control a matching receiver that is placed anywhere within the active area of the beam. Note that the Tx and Rx do not need to be pointed directly at each other to give effective operation, but *must* be in line-of-sight contact; also note that an object placed within the beam can create a blind area in which line-of-sight contact cannot exist.

Optoreflectors

Introduction

An optoreflector is a substance or device that has an ability to usefully reflect visible or IR or UV light. Many natural substances, such as fog, smoke, and various liquids, are significantly optoreflective, and their presence can thus easily be detected via fairly simple optoelectronic circuitry. Man-made objects such as mirrors and bright-metal surfaces are highly reflective, and have many applications in optoelectronics, but one of the most useful of all man-made optoreflectors is the liquid crystal display, or LCD, which is described in some detail in this section.

Liquid crystal displays (LCDs)

Optoelectronic display devices such as LEDs inevitably consume significant electric power, since they directly *generate* light. Liquid crystal displays (LCDs), on the other hand, selectively *reflect* existing ambient light, and can thus operate with negligible power consumption. Liquid crystals are mesomorphic (intermediate between a solid and a liquid) organic substances; the types used in LCDs are known as nematic (threadlike), and can have their light-transmission characteristics temporarily changed by the application of a low-voltage external electric field.

Figure 1.24 shows the basic construction of an LCD designed to display either a blank or the symbol 1. The device consists of a thin layer of liquid crystal, sandwiched between two clear glass covers and sealed in place by a plastic gasket. The upper (front) cover has a mirror-image of the symbol '1' printed on its *rear* face in transparent electrically conductive film; this symbol is electrically connected to a 'back plane' (BP) terminal. The lower cover has the symbol '1' printed on its *upper* face in transparent electrically conductive film and is electrically connected to an 'S' terminal, and has an opaque coloured film on its rear.

During manufacture, the above components are aligned and sealed together so that the upper and lower printed symbols are precisely opposite each other (aligned); note that under this condition the printed electrical

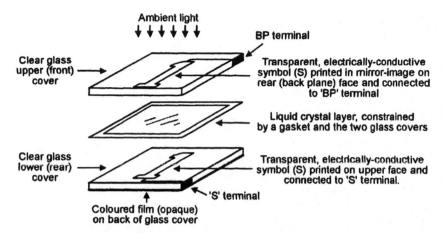

Figure 1.24 *Exploded view showing basic elements of a simple LCD device designed to display either a blank or the symbol 1*

connection lines of the upper and lower symbols are *not* aligned.

The basic device action is such that the liquid crystal film is normally fully transparent, and the rear opaque coloured film is thus clearly visible under this condition. When, however, a symmetrical a.c. squarewave is applied between the BP and S terminals, the localized area of liquid crystal *between the aligned upper and lower symbols* changes its crystalline alignments and becomes clearly visible – in the shape of the selected symbol superimposed on the rear opaque background colour – through the LCD's upper cover; the symbol's electrical connections – which are not aligned – are not visible under this condition.

The first generation of LCDs, developed during the 1970s, used a *dynamic scattering* type of nematic liquid crystal, in a 50mm thick film, in which the rear opaque background was usually black and in which the area of liquid crystal between the selected upper and lower symbols took on a mirror-like quality when the LCD was stimulated by an external a.c. energizing voltage. This type of display – which characteristically produced a silvery-white display on a black background – proved to be very unreliable, often having a working life of only 1000 hours, and is now rarely (if ever) used.

The present generation of LCDs use a *twisted nematic* type of liquid crystal, which has the strange property – under normal conditions – of linearly rotating the plane of polarization of light as it passes through the crystal; the amount of rotation is proportional to the thickness of the liquid crystal, at a typical rate of 90° per 10µm of thickness, but falls to 0° when an electric field is applied across the liquid crystal.

Most modern LCDs use the basic type of construction shown in *Figure 1.25*. Here, a 10µm film (equal to 1/5th of the thickness of a human hair) of

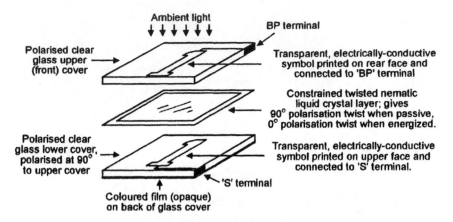

Ambient light

BP terminal

Polarised clear
glass upper
(front) cover

Transparent, electrically-conductive
symbol printed on rear face and
connected to 'BP' terminal

Constrained twisted nematic
liquid crystal layer; gives
90° polarisation twist when passive,
0° polarisation twist when energized.

Polarised clear
glass lower cover,
polarised at 90°
to upper cover

Transparent, electrically-conductive
symbol printed on upper face and
connected to 'S' terminal.

'S' terminal

Coloured film (opaque)
on back of glass cover

Figure 1.25   *Exploded view showing basic elements of a modern 'twisted nematic' LCD device designed to display either a blank or the symbol 1*

twisted nematic liquid crystal is sandwiched between two layers of *polarized* glass, which have their polarization fields aligned at right angles to each other. Note that when two polarization fields are aligned in the same plane, the combination is optically transparent, but when they are aligned at right angles the combination is opaque. Thus, the action of the *Figure 1.25* LCD device is as follows:

The basic device action is such that, under 'passive' conditions, external polarized light passes through the front glass, is twisted by 90 degrees as it passes through the liquid crystal film, and thus emerges in the same polarization plane as the rear glass and hence makes contact with the coloured film (normally silver) on the rear of that glass, which is thus clearly visible under this condition.

When, on the other hand, a symmetrical a.c. squarewave is applied between the BP and S terminals, the area of liquid crystal between the selected '1' symbols ceases to twist polarized light, and under this condition the incoming polarized light is completely blocked (absorbed) as it reaches the opposing polarization field below the '1' symbol. Thus, under this condition the selected symbol becomes clearly visible – as a black symbol superimposed on the rear opaque background colour (usually silver) – through the LCD's upper cover.

Note that LCDs degrade very rapidly if energized by a d.c. source, and consequently *must* be energized (driven) by a perfectly symmetrical a.c. source (usually a squarewave) with zero d.c. components. The required type of waveform is usually derived from a basic DC supply via a simple voltage-doubling 'bridge-drive' circuit, which takes the basic form shown in *Figure 1.26*.

In *Figure 1.26*, a symmetrical squarewave is applied to the input of a series-connected pair of CMOS inverters, which are shown powered from a 10V DC source and thus each generates a 10V peak-to-peak squarewave output,

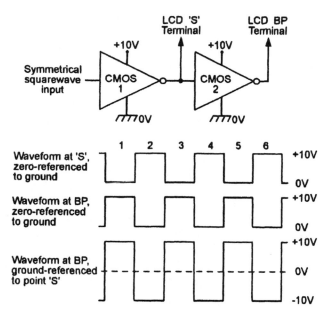

Figure 1.26 *Basic circuit and waveforms of a voltage-doubling 'bridge-drive' circuit suitable for driving a liquid crystal display*

which is fed to LCD terminal S or BP; note that these two outputs are in anti-phase. Thus, during period 1 of the drive signal, terminal BP is 10V positive to terminal S and is thus seen as being at +10V. In period 2, however, terminal PB is 10V negative to terminal S, and is thus seen as being at −10V. Consequently, if terminal S is regarded as a zero voltage reference point, it can be seen that the terminal-BP voltage varies from +10V to −10V between period 1 and 2; this waveform is perfectly symmetrical and has zero d.c. components, and is thus ideal for driving an LCD.

Finally, note that one of the many advantages of the LCD is that (since its symbols take a printed form) its symbols can take any desired shape, and are limited in number only by the size of the display and by the need to make external electrical connections to each symbol. In alphanumeric displays (such as used in clocks, calculators, and digital instruments) the symbols are usually arranged in 7-segment or dot-matrix format. Basic details of these display formats are given in the following, concluding, section of this chapter.

LED/LCD display formats

Light-emitting diodes (LEDs) have very simple drive requirements, and consequently are often used – individually or in small groups – as simple

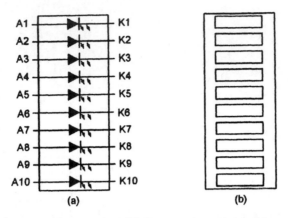

Figure 1.27 *Basic circuit of a 10-LED bar-graph package (a), and (b) typical appearance of the device's frontal (display) aspect*

attention-grabbing 'indicators' that give a steady or a flashing visual display, either to indicate the presence or absence of a voltage or signal, or purely for amusement (as in electronic jewelry). Chapter 2 describes a variety of simple LED circuits of these types.

LEDS can also be used – in either individual or special-purpose multi-LED packaged form – as electronic 'chasers' or sequencers. These are electronic circuits that drive an array of LEDs in such a way that individual LEDs (or groups of LEDs) turn on and off in a predetermined and repeating sequence, thus producing a visually attractive display in which a ripple of light seems to run along a chain. Chapter 3 presents a small selection of practical chaser/sequencer circuits.

One widely used type of multi-LED display is the bar-graph type, in which the display often takes the form of a dedicated multi-LED package. *Figure 1.27(a)* shows the basic circuit of a 10-LED bar-graph package (often called a '10-segment' display), and *Figure 1.27(b)* shows the typical appearance of the device's frontal (display) aspect. Displays of this type can usually be stacked end-to-end, to give a display of any desired length.

Bar-graph packages are usually used to indicate the analogue value of a voltage or signal, in a similar manner to an analogue voltmeter. They are driven by a special IC that activates a line (chain) of LEDs, the length of the line being directly proportional to the magnitude of the analogue input signal. If the line of LEDs is illuminated like a solid bar, the display is known as a 'bar-graph' display; if only a single LED is illuminated at the end of the line, the display is known as a 'dot-graph' display. Chapter 4 presents circuits of both of these types.

One of the most widely used LED or LCD displays is the type that can show alphanumeric characters: digital watches, pocket calculators, and digital instruments are all examples of devices that use such displays. The best-

Figure 1.28 *Basic format of a 7-segment display*

known display of this type is the 7-segment display, which – in its most basic form – comprises seven independently accessible photoelectric segments (usually LED or LCD types) arranged in the form shown in *Figure 1.28*.

In these displays, the seven display segments are conventionally notated from *a* to *g* in the manner shown in *Figure 1.28*, and it is possible to make them display any numeral from 0 to 9 or any alphabetic character from A to F (in a mixture of upper and lower case letters) by activating these segments in various combinations. *Figure 1.29* shows the appearance of the numbers 0 to 9 when illuminated in this way.

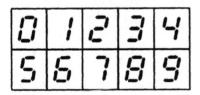

Figure 1.29 *The numbers 0 to 9 presented in 7-segment display format*

In LED-type 7-segment displays, all the anodes (or cathodes) of the LEDs are usually joined together (made common) internally, and the display is then known as a 'common-anode' (or 'common-cathode') type, and requires the use of a minimum of only eight external control terminals. In practice, most such displays incorporate an additional LED, which is used as a decimal point (DP) display and is usually placed to the lower-right of the main display. Most 7-segment displays are reasonably easy to use and can be activated via a variety of dedicated decoder/driver ICs. Chapter 5 deals with this subject in depth.

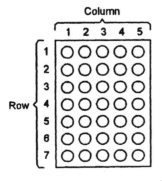

Figure 1.30 *Basic 5 × 7 dot-matrix display structure*

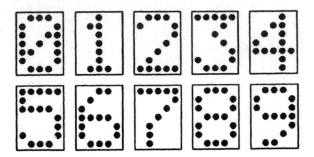

Figure 1.31 *Typical appearances of the numbers 0 to 9 in 5 × 7 dot-matrix display format*

For more sophisticated alphanumeric displays, dot-matrix devices are often used. The most popular of these use the 5 × 7 dot matrix display type of structure shown in *Figure 1.30*, which houses 35 individual 'dot' display elements (usually LEDs), each of which can be individually accessed by selecting an appropriate column/row combination. Thus, the upper L/H dot can be accessed via column 1, row 1; the central dot can be accessed via column 3, row 4; the lower R/H dot can be accessed via column 5, row 7, and so on.

Simple 5 × 7 dot matrix units can – by using multiplexing techniques – be used to display any normal character in the standard ASCII alphanumeric character set; *Figure 1.31*, for example, shows the typical appearance of the numbers 0 to 9 when displayed in 5 × 7 dot matrix format.

Dot-matrix displays are not easy to use; they require drive circuitry that is far more sophisticated than that used to drive 7-segment displays. Because of this problem, many dot-matrix displays are produced in 'smart' or 'semi-intelligent' form, with built-in ROM, code-conversion and drive circuitry that can be fed directly from the output of a microprocessor or an ASCII code generator system. They can thus be regarded as special-purpose (rather than general-purpose) optoelectronic display devices.

2

Basic LED display circuits

The simple LED (light-emitting diode) is the most important and widely used of all modern optoelectronic devices. It emits a fairly narrow spectrum of visible (usually red, orange, yellow or green) or invisible (infra-red) light when its internal diode junction is stimulated by a forward electric current, has good power-to-light energy conversion efficiency, and has a very fast response time (often less than 0.1µs). This chapter explains LED operating principles and characteristics and shows a variety of ways of using LEDs as basic visual indicators, as simple 'flashing light' displays, and as micropower flashing lights.

LED basics

Introduction

Figure 2.1(a) shows the standard symbol used to represent a LED in this volume, together with its anode (a) and cathode (k) terminal notations. Normal LEDs are *pn* junction diodes and use the basic semiconductor structure shown in *Figure 2.1(b)*, but are made from exotic semiconductor materials such as gallium arsenide phosphide (GaAsP) or aluminium–gallium

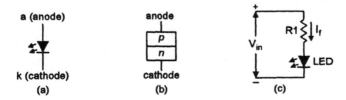

Figure 2.1 *Standard LED symbol and terminal notations (a), basic semiconductor structure (b), and basic way of using a LED as a simple light generator (c)*

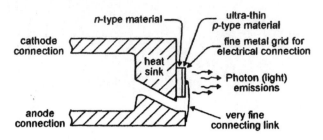

Figure 2.2 *Typical internal structure of a normal LED*

arsenide (AlGaAs), etc. Like normal junction diodes, they pass current easily in the forward direction but block it in the reverse direction.

Figure 2.1(c) shows the basic way of using a LED as a simple light generator, powered from a fixed-voltage supply. Resistor R1 limits the LED's forward current (and thus the LED brightness) to a sensible value. Normal LEDs operate in the basic way described in the next paragraph.

When a *pn* junction diode is conducting in the forward direction, it releases energy as its charge carriers (electrons and holes) recombine as the electrons jump from the semiconductor's conduction band to its valence band. In a normal silicon diode, most of this energy is released in the form of heat, plus a minute amount of radiant (light) energy in the form of photons. In a LED, however, the released radiant energy is – because of the special and exotic nature of the device's semiconductor material – several orders of magnitude greater than in a normal silicon diode, and LEDs thus act as excellent generators of visible or IR light. The generated light's wavelength (and thus its colour) depends on the precise formulation of the semiconductor material.

Figure 2.2 shows the typical internal structure of a normal LED. The *p* and *n* materials are fused together, with the *n* material bonded to a heat sink that also forms the cathode connection. The *p* material is ultra-thin and translucent, and is connected to the anode terminal via a very fine bonded-on metal grid and a fine wire link. When the LED is active, photons are generated at the *pn* junction and radiate outwards, through the surface of the translucent *p* layer and the metal grid, and finally emerge from the 'viewing' end of the LED. This standard type of LED is known (for obvious reasons) as a surface-emitting LED; more-complex edge-emitting LEDs are also manufactured for use in special data-transfer applications.

When a surface-emitting LED is active, its semiconductor emitting surface is uniformly bright and the light radiates in a cone-like manner, with its strength greatest at right angles to the surface and diminishing in strength at declining angles. This light radiation pattern is known as *Lambertian* and takes the form shown in *Figure 2.3*; the power levels fall to 50% of maximum at the 60° viewing angle points, i.e. the pattern has a half-power bandwidth

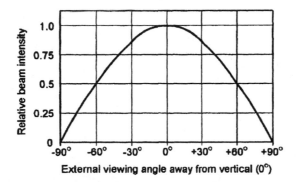

Figure 2.3 *Lambertian radiation pattern produced from the semiconductor surface of a normal LED. The half-power width is 120°*

of 120°. Most 'round' LEDs incorporate a simple lens that modifies the light pattern actually radiated by the LED.

Roughly 2V are developed across most LEDs when passing a useful forward current; *Figure 2.4* lists the typical forward volt drops (V_f) of different coloured standard 5mm diameter LEDs at forward currents of 20mA; the actual voltage value varies with the current value. If a LED is reverse biased it may avalanche or zener at a fairly low voltage value, typically less than 12V.

LEDs are available in a variety of styles, the most popular being the 'round' type that has the basic shape shown in *Figure 2.5* and which is made in standard diameter sizes of 3mm, 5mm (0.2in), 8mm, or 10mm. Round LEDs use a clear or coloured plastic case with a lens moulded into its dome, and are designed to be viewed end-on, looking towards the dome, as indicated in the diagram. The LED case has a polarity-identifying flat moulded into the side of its base adjacent to the cathode lead, which is usually shorter than the anode lead when untrimmed. Special fittings are readily available for fixing most sizes of LED to front panels, etc.

One important but confusingly-named LED parameter is its 'viewing angle', at the extremes of which the LED's optical output intensity falls to half of its maximum axial value. Some LEDs give a diffused output, in which the light intensity falls off gradually beyond the viewing angle and is thus

Colour	Red	Orange	Yellow	Green	Blue
V_f (typical)	2V	2V	2.1V	2.2V	3.3V

Figure 2.4 *Typical forward voltage values of standard LEDs at a current-limited value of 20mA*

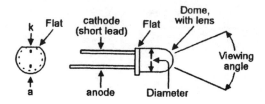

Figure 2.5 *Typical physical details of 'round' LEDs and methods of recognizing their polarity*

clearly discernible over a wide angular range; others (particularly Hyper-bright types) have a sharply focused output in which the light intensity falls off very sharply beyond the specified viewing angle.

LEDs are available in five different brightness categories, which are usually known as Standard, High Brightness, Super Bright, Ultrabright, and Hyperbright. LED brightness level is almost linearly proportional to the operating current value over most of the LED's normal (DC) operating current range. The brightness is usually specified in millicandelas (mcd), with the LED passing an operating current of 20mA. The table of *Figure 2.6* presents typical optical output power and viewing angle figures for the five types of 5mm round LED. Note in the Red LED column that the Ultrabright and Hyperbright devices (which use water-clear lenses) are 143 and 500 times brighter, respectively, than a standard red LED.

LED type	Viewing angle	Red	Green	Orange
Standard	60°	7mcd	5.2mcd	8mcd
High Brightness	40°	30mcd	25mcd	50mcd
Super Bright	30°	125mcd	120mcd	140mcd
Ultrabright	25°	1000mcd	——	——
Hyperbright	25°	3500mcd	——	——

Figure 2.6 *Typical viewing angle and optical output power figures (in millicandelas) of five basic types of 5mm red, yellow and green LEDs*

LED jargon and data

When designing LED-based circuits, it is useful to be able to decipher the jargon used in the manufacturer's data sheets. The two most important sets of jargon are those that relate to the LED's light output characteristics and those that relate to the LED's input-to-output power characteristics. *Figures 2.7 to 2.11* help explain the most important pieces of this jargon.

All man-made light generators produce a spectrum of light wavelengths rather than a single wavelength. In tungsten lamps, the spectrum is very wide; in lasers, it is very narrow; in LEDs the spectrum is moderately narrow.

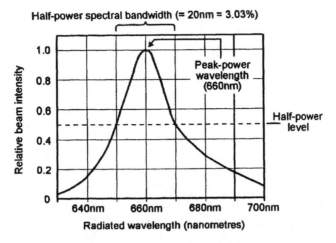

Figure 2.7 *Typical light output spectrum of a red LED*

Figure 2.7 shows the typical light output spectrum of a red LED, which in this example has a peak-power wavelength of 660nm; other typical LED peak-power wavelengths are 565nm for green, 595nm for yellow, 610nm for orange, and 880nm to 950nm for IR LEDs.

The LED's spectral bandwidth (sometimes called *linewidth*) is measured between the two half-power radiation (output intensity) levels and may be expressed in actual wavelength variation (20nm in this example) or as that variation expressed as a percentage of the peak-power wavelength (as 3.03% in this case).

In applications where a LED is required to act as a simple or flashing indicator, vital data such as its operating current and power level limits, etc., can be obtained from manufacturer's or retailer's basic data tables. *Figure 2.8* shows a typical data table for a range of 5mm Standard diffused red, orange, yellow and green LEDs. Note that the listed maximum power and

	Red	Orange	Yellow	Green
I_f (typ.)	10mA	10mA	10mA	10mA
I_f (max)	30mA	30mA	25mA	30mA
V_f (typ.)	2.0V	2.0V	2.1V	2.2V
V_R (max)	5V	5V	5V	5V
Intensity (typ., at 10mA)	4mcd	14mcd	15mcd	10mcd
Viewing angle (typ.)	60°	60°	60°	60°
Peak wavelength	660nm	610nm	595nm	565nm
Power dissipation (max)	130mW	120mW	100mW	120mW

Figure 2.8 *Typical data table for a range of 5mm Standard diffused LEDs*

current values are DC-equivalent (rather than instantaneous absolute or pulse) values; the LED may be damaged if these DC values are exceeded.

A LED's operating current and power limits are determined mainly by the thermal characteristics of its very small active *pn* area. If a red LED is powered from a DC source, about 60mW are generated in its *pn* area at an operating current of 30mA, and may typically raise the temperature of its *pn* area to 55°C. At 60mA, the generated power rises to about 150mW and the *pn* area's temperature reaches its maximum operating limit of 100°C; any further rise in operating current (or power) may destroy the *pn* area.

In practice, the physical mass of the LED's *pn* area and its heat sink (see *Figure 2.2*) provide it with thermal inertia, and the *pn* area thus takes a finite time to heat up in response to sudden power surges. The relationship between surge current and response time is normally given in the LED's expanded data sheet, under the *Absolute maximum ratings* section, and is noted in terms such as '1A <10µs', and in this example means that the LED can safely handle a 1A current pulse for a maximum of only 10µs.

Note that the above '1A <10µs' value gives a direct relationship between permitted current and pulse duration values and can be safely translated into alternative values at currents *less* than 1A, thus giving values such as '100mA <100µs', or '50mA <200µs', etc. This data applies to non-repetitive pulses; if the pulses are repetitive, their *mean* power dissipation (integrated over one complete repetition period) must not exceed the LED's normal DC operating power limit.

The above data is of great value when designing LED remote control and remote signalling systems, in which the transmitter LED is energized by repetitive high-value current pulses. In such designs, additional basic design data must be derived from the manufacturer's expanded data sheets (available directly from the manufacturer or via their Web site on the Internet). *Figures 2.9* to *2.11* show typical examples of such data sheets.

Figure 2.9 shows the typical relationship between luminous output intensity (normalized to unity at a forward current of 20mA) and forward current over a LED's normal 0–30mA operating range; the intensity is almost linearly proportional to forward current at values up to 20mA.

Figure 2.10 shows the typical relationship between the above LED's luminous output intensity (normalized to unity at a forward current of 20mA) and the LED's forward current value over the extended 10mA to 1A operating range (this particular LED has a DC forward current limit of 50mA, a surge current limit value of 1A <10µs, and a DC power limit of 150mW). Note that the normalized light intensity rises to a peak value of 1.6 at a forward current of 100mA but decreases at higher currents. The highest useful pulse current is thus 100mA, at which the maximum permitted pulse length is thus 100µs.

Figure 2.11 shows the typical relationship between the above LED's forward voltage and its forward current, over the extended operating range.

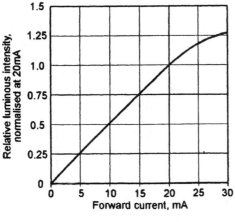

Figure 2.9 *Typical relative luminous intensity versus forward current graph covering normal operating ranges*

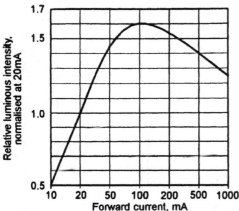

Figure 2.10 *Typical relative luminous intensity versus forward current graph covering extended (pulse) operating ranges*

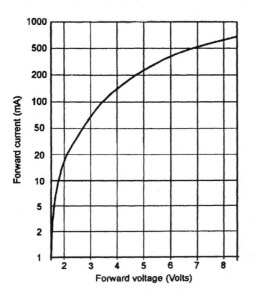

Figure 2.11 *Typical forward voltage/current graph covering extended (pulse) operating ranges*

Note that this particular LED generates a forward voltage of 3.5V when fed with a 100µs 100mA pulse, and its *pn* junction thus receives a peak power input of 350mW under this condition. If this pulse is applied in the form of a symmetrical squarewave, the mean values of both the current and the voltage are halved, and the *mean* power value is thus reduced by a factor of four, to 87.5mW in this case; this value is well within the safe operating limits of this particular LED.

Thus, the above data shows that this particular LED can be efficiently and safely used as a remote-control or signalling transmitter by feeding it with controlled squarewave signals with a peak value of 100mA and a minimum operating frequency of 5kHz (i.e. with a maximum cycle period of 200µs).

Basic LED indicator circuits

A LED can be used as a simple light-generating 'indicator' by wiring it in series with a current-limiting resistor and connecting the combination across a suitable DC supply voltage. *Figure 2.12* shows how to work out the resistance (R) value needed to give a particular current from a particular DC supply voltage. Thus, if a red LED has a forward volt drop of 2V at 20mA and is required to operate at 20mA from a 10V supply, R needs a value of $(10V - 2V)/0.02A = 400R$. In practice, R can be connected to either the anode or the cathode of the LED.

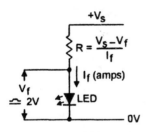

Figure 2.12 *Method of finding the R value for a given V_S and I_f*

A LED can be used as an indicator in an AC circuit by wiring it in any one of the three alternative ways shown in *Figure 2.13*. In *Figure 2.13(a)* the LED is wired in inverse parallel with a IN4148 (or similar) silicon diode, to prevent the LED being reverse biased. The LED is fed with half-wave current in this mode, with only half of R1's current flowing through the LED; for a given LED brightness, the 'R' value must be halved relative to that indicated in the *Figure 2.12* DC circuit. Thus, to operate the LED at a mean current of 10mA from a 100V AC supply, this design needs an R1 value of about 5k0, and R1 needs a power rating of at least 2 watts.

In *Figure 2.13(b)* the LED is wired in series with rectifier D1 (which must have a reverse-voltage rating greater than the peak voltage value of the AC

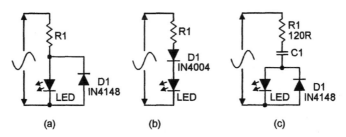

Figure 2.13 *Ways of using a LED as an indicator in an AC circuit*

supply), to protect the LED when it is reverse biased. The LED is fed with half-wave current in this mode, with all of this current flowing through R1, and for a given LED brightness the 'R' value must again be halved relative to that indicated in the *Figure 2.12* circuit. Note in this circuit, however, R1's mean current and voltage values are both halved relative to those of the *Figure 2.13(a)* design. Thus, to operate the LED at a mean current of 10mA from a 100V AC supply, this design needs an R1 value of about 5k0, and R1 needs a power rating of at least 0.5 watts.

In *Figure 2.13(c)* the LED is wired in inverse parallel with silicon diode D1, and the LED's mean current is determined mainly by the values of the AC supply voltage and the impedance of capacitor C1 (R1 simply limits switch-on surge currents to a safe value). Note that, since a capacitor's voltage and current are 90° out of phase, negligible power is dissipated in C1 in this circuit, which is usually used at 110V to 240V AC power line voltages. When powered by a 50–60Hz AC supply, a C1 value of 0.47μF gives a mean LED current of about 10mA from a 120V supply, or 20mA from a 240V supply; the current value is directly proportional to the C1 value.

Special-purpose LEDs

LEDs are readily available in a variety of special-purpose forms, the best known of which are the 'direct connection' type, the 'flasher' type, and the multi-colour types.

Direct connection LEDs are designed to be connected directly across a fixed-value DC or AC voltage source. DC voltage types take the basic form shown in *Figure 2.14(a)* and incorporate a current-limiting resistor that is

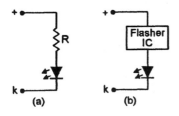

Figure 2.14 *Basic forms of a direct connection DC LED (a) and a flasher LED (b)*

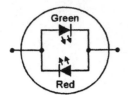

Figure 2.15 *Bi-colour LED actually houses two LEDs connected in inverse parallel*

housed in the LED body in 5V and 12V types, or in one of the LED leads in higher voltage types. AC voltage types (usually designed for use with 110V or 240V supplies) take one of the basic forms shown in *Figure 2.13* but are usually housed in an insulated panel-mounting assembly.

Flashing LEDs take the basic form shown in *Figure 2.14(b)* and have a built-in integrated circuit that gives the flashing effect. They are available in red, green and yellow, have a typical flashing frequency of 2Hz, and can (typically) use 6V to 12V DC supplies.

Multi-colour LEDs are actually 2-LED devices; *Figure 2.15* shows a bi-colour device that comprises a red and a green LED connected in inverse parallel, so that the colour green is generated when the device is connected in one polarity, and red is generated when it is connected in the reverse polarity. This device is useful as a polarity or null indicator.

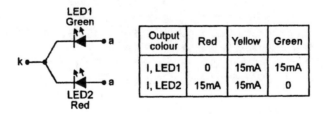

Output colour	Red	Yellow	Green
I, LED1	0	15mA	15mA
I, LED2	15mA	15mA	0

Figure 2.16 *Multi-colour LED, giving three colours from two junctions*

Figure 2.16 shows another type of multi-colour LED, which is sometimes known as a tri-colour type. This comprises a green and red LED mounted in a 3-pin common-cathode package. This device can generate green or red colours by turning on only one LED at a time, yellow by turning both LEDs on by equal amounts, or any colour between green and red by turning both LEDs on in the appropriate ratios.

Multi-LED circuits

If several LEDs need to be driven from a single power source, this can be done by wiring all LEDs in series, as shown in *Figure 2.17*, provided that the supply voltage is significantly greater than the sum of the individual LED forward voltages. This circuit thus consumes a minimum total current, but is

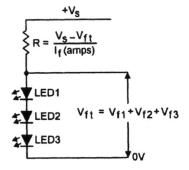

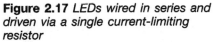

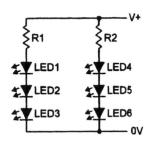

Figure 2.17 *LEDs wired in series and driven via a single current-limiting resistor*

Figure 2.18 *Any number of Figure 2.17 circuits can be wired in parallel, to drive any number of LEDs*

limited in the number of LEDs that it can drive. Any number of these basic circuits can, however, be wired in parallel, so that any number of LEDs can be driven from a single source, as shown in the 6-LED circuit of *Figure 2.18*.

An alternative way of simultaneously powering several LEDs is to simply wire a number of the *Figure 2.12* circuits in parallel, as shown in *Figure 2.19*; note, however, that this circuit is very wasteful of supply current (which equals the sum of the individual LED currents).

Figure 2.20 shows a 'what *not* to do' LED-driving circuit, in which all the LEDs are wired directly in parallel. Often, this circuit will not work correctly because inevitable differences in the forward characteristics of the LEDs causes one LED to hog most or all of the available current, leaving little or none for the remaining LEDs.

LED flasher circuits

Simple designs

One of the simplest types of LED display circuit is the LED flasher, in which a single LED repeatedly switches on and off, usually at a rate of one or two flashes

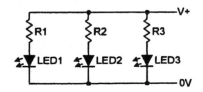

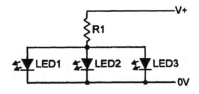

Figure 2.19 *This circuit can drive any number of LEDs, but at the expense of high current*

Figure 2.20 *This LED-driving circuit may not work; one LED may hog most of the current*

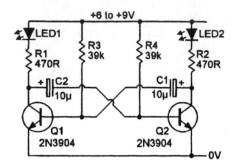

Figure 2.21 *Transistor 2-LED flasher circuit operates at about 1Hz*

per second. A 2-LED flasher is a simple modification of this circuit, but is arranged so that one LED switches on when the other switches off, or vice versa.

Figure 2.21 shows the practical circuit of a transistor 2-LED flasher, which can be converted to single-LED operation by simply replacing the unwanted LED with a short-circuit. Here, Q1 and Q2 are wired as a simple 1Hz astable multivibrator, in which Q1 and LED1 turn on as Q2 and LED2 turn off, and vice versa, and in which the astable switching rates are controlled by the values of C1–R3 and C2–R4.

Figure 2.22 shows an IC version of the 2-LED flasher, based on a 555 or 7555 timer IC that is wired in the astable mode, with its main time constants determined by the C1 and R4 values and giving a cycling rate of about 1Hz (one flash per second). The circuit action is such that output pin 3 of the IC alternately switches between the ground and the positive supply voltage levels, alternately pulling LED1 on via R1 or driving LED2 on via R2. The circuit can be converted to single-LED operation by omitting LED2 and R2.

Figure 2.23 shows a useful modification of the above circuit, in which the flashing rate is made variable via RV1, and two pairs of series-connected LEDs are connected in the form of a cross so that the visual display alternately switches between a horizontal bar (LED1 and LED2 ON) and a vertical bar (LED3 and LED4 ON), thus forming a visually interesting display. The cycling rate is variable from 0.3 to 3 flashes per second.

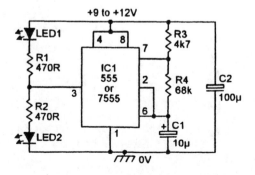

Figure 2.22 *IC 2-LED flasher circuit operates at about 1Hz*

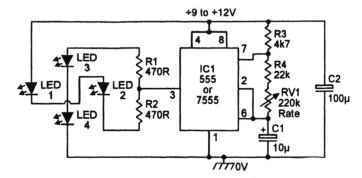

Figure 2.23 *The rate of this 4-LED double-bar flasher is variable from 3 to 0.3 flashes per second*

Micropower LED flashers

Simple LED flasher circuits of the types shown in *Figures 2.21* to *2.23* consume mean operating currents of several milliamps. Micropower LED flashers, on the other hand, consume mean operating currents that are measured in microamps (typically ranging from 2µA to 150µA), and are intended mainly for use in battery powered 'emergency indicator', 'battery state', and 'burglar deterrent' applications.

In emergency indicator applications, micropower LED flashers can be used to indicate the positions of emergency exits, lanterns, torches, alarm buttons, or safety equipment, etc., under dark conditions (perhaps caused by a failure of a main lighting system). When used as battery state indicators, they are often fitted in smoke alarms and other low-current long-life units that are powered by 4.5V to 12V batteries. When used as burglar deterrents, they are prominently fitted to real or dummy burglar alarm control or alarm/siren boxes or CCTV cameras, etc.

To understand the basic principles behind micropower LED flashers you must first learn some basic facts concerning visual perception, as follows.

(1) The human eye/brain combination is sharply attracted by sudden changes in visual patterns or light levels; it is particularly sensitive to some types of flashing light. *Figure 2.24* shows the typical 'light flash' response of the human eye/brain combination when presented with a bright LED-generated pulse of light.
(2) Note from *Figure 2.24* that the flash must be present for at least 10ms to be seen (perceived) at full brilliance, and that – when the flash terminates – the 'persistence of vision' effect causes the perceived brilliance to decay fairly slowly, typically taking 20ms to fall to 50% of its maximum (pre switch-off) value. Consequently, the eye can only see

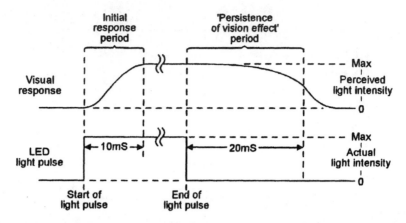

Figure 2.24 *Typical 'light flash' response of the human eye/brain combination*

flashing lights as *individual* flashes if they are separated by a period of at least 20ms; if they are separated by less than 20ms they are seen (because of the 'persistence of vision' effect) as a continuous light.

(3) Also note from *Figure 2.24* that, if the flashes are separated by at least 20ms, the brain 'sees' the individual flashes at full brilliance if they have a duration of 10ms or greater, but sees them at diminishing brilliance at durations below 10ms (a 2ms flash appears at roughly 1/5th of true brilliance; the perceived brilliance falls off rapidly at durations below 1ms). The perceived *duration* of a 20ms flash (30ms) is only 50% greater than that of a 10ms flash (20ms).

(4) The human eye/brain combination is very sharply attracted by flashing lights that have repetition periods in the approximate range 0.5 to 5 seconds, but is less attracted by flashing lights that have repetition periods above or below this range.

(5) Modern low-cost Super Bright LEDs, when generating a 10ms or longer light pulse, produce a brightness level that is adequately eye-catching for most practical purposes when pulsed by a 2mA current.

When the above sets of facts are put together, it transpires that the 'ideal' micropower LED flasher (when using a Super Bright LED) should produce a pulse with a duration (d) of 10ms at a current (I) of 2mA, at a repetition period (p) of 2 seconds (= 2000ms). Note that, under these conditions, the *mean* current (I_{mean}) of the LED is given by

$$I_{mean} = I \times d/p$$

and is a mere 10µA in this particular example (at a 30 second repetition period, I_{mean} is a minute 0.67µA).

V+	R3	Current drain	
		Loaded	No Load
3V	1.5k	68μA	60μA
4.5V	1.5k	80μA	70μA
6V	2.2k	86μA	75μA
9V	3.3k	97μA	85μA
12V	4.7k	107μA	95μA

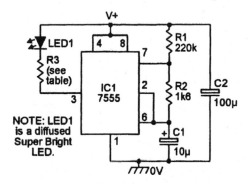

Figure 2.25 *Circuit and performance details of a 7555–based micropower LED flasher unit*

In practice, the *actual* mean current consumed by a micropower LED flasher circuit is equal to the sum of the LED and the driver currents, and is inevitably higher than the minimum figure indicated above. *Figures 2.25* and *2.26*, for example, show two alternative micropower LED flasher circuits that, when powered from 6V supplies, consume total currents of 86μA and 12μA respectively.

The *Figure 2.25* circuit is designed around a CMOS 7555 timer IC that is used in the astable mode and typically consumes an unloaded operating current of 75μA at 6V. In this mode, C1 alternately charges up via R1–R2 and discharges via R2 only, thus generating a highly asymmetrical output waveform on pin-3, which pulls the LED on via current-limiting resistor R3 during the brief discharge part of each operating cycle. The *Figure 2.25* table summarizes the circuit's performance details when optimized for operation at various spot voltages in the range 3V to 12V.

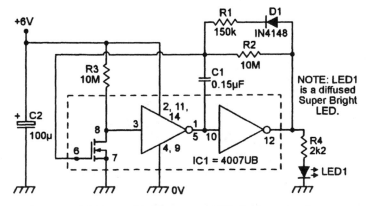

Figure 2.26 *This 4007UB-based micropower LED flasher circuit consumes a mean current of 12μA at 6V*

The *Figure 2.26* circuit is designed around a CMOS 4007UB IC, which contains two complementary MOSFET transistor pairs plus one CMOS inverter, all housed in a 14-pin DIL package. In this application the IC is wired as a micropower ring-of-three asymmetrical astable multivibrator which, when powered from a 6V supply, drives the LED on for 10ms at 2–second repetition intervals; the ON time is controlled by C1–R1, the OFF time by C1–R2, and the LED current (2mA nominal) is controlled by R4.

C2 is a vital part of the *Figure 2.26* circuit and acts as a low-impedance energy store that enables the circuit to work well even when powered from a fairly high-impedance source such as a weak battery. The complete circuit consumes an unloaded operating current of 2µA, and a loaded current (when driving the LED with 2mA pulses) of 12µA.

Note that the *basic* circuit of *Figure 2.26* can be used at any supply voltages in the range 4.5V to 12V, but that the actual component values must be selected to suit the specific supply voltage used. Also note that, at supply voltages of 6V or greater, the circuit can drive two or more series-connected LEDs without increasing the total current consumption, provided that R4's value is altered to set the LED ON currents at 2mA.

The table of *Figure 2.27* shows the nominal life expectancies of various types of alkaline cell/battery when continuously driving various types of micropower LED flasher circuit. The data relate to the circuits of *Figures 2.25* (drawing 86µA at 6V) and *2.26* (drawing 12µA at 6V), and to the once-popular but now obsolete LM3909 'LED flasher' IC (described in the 1st edition of this book but withdrawn from production by National Semiconductor in 1998/9), which draws a minimum operating current of 320µA.

Note in *Figure 2.27* that the 'predicted cell/battery life' figures relate to cells/batteries that have initial (unused) life expectancies of five years, i.e. in which their charges leak away at a steady rate of 1.67% per month. The total in-use monthly capacity drain equals the sum of the leakage and the loading drain figures, and forms the basis of the life prediction figures shown in the table.

Alkaline cell/battery type	Capacity (per cell or battery)	12µA load		86µA load		320µA load	
		monthly capacity drain & predicted cell/battery life					
		drain	life	drain	life	drain	life
AAA (1.5V)	1Ah	0.88%	3.3 Yrs	6.28%	1.0 Yrs	23.4%	0.3 Yrs
AA (1.5V)	2Ah	0.44%	4.0 Yrs	3.14%	1.7 Yrs	11.7%	0.6 Yrs
C (1.5V)	6.5Ah	0.135%	4.6 Yrs	0.97%	3.2 Yrs	3.6%	1.6 Yrs
D (1.5V)	13Ah	0.07%	4.8 Yrs	0.48%	3.9 Yrs	1.8%	2.4 Yrs
PP3 (9V)	0.55Ah	1.59%	2.6 Yrs	11.4%	0.6 Yrs	42.5%	0.2 Yrs

Figure 2.27 *Table showing the life expectancies of various types of alkaline cell/battery when driving micropower LED flasher circuits*

Low-voltage micropower LED flashers

The basic micropower LED flasher circuit of *Figure 2.26* can, if its component values are suitably selected, be reliably used at an absolute minimum supply voltage of 4.5V. Consequently, if you want to drive a micropower LED flasher from a 3V battery and do not want to use the 3V version of the 7555–based *Figure 2.25* circuit, one simple option is to drive the 6V *Figure 2.26* circuit from the 3V battery via a voltage-doubler circuit based on an ICL7660 IC. *Figure 2.28* shows the complete circuit of a micropower LED flasher unit of this type, which typically draws a mean operating current of about 35μA from the 3V battery.

In *Figure 2.28*, 1C2 incorporates a free-running symmetrical squarewave generator that drives a ganged pair of CMOS bilateral switches that have their 'common' terminals connected to the two ends of C3 via pins 2 and 4. In the first half of each cycle, these switches connect pin-4 to the pin-3 0V line and connect pin-2 to the pin-8 '+3V' line, thus charging C3 up to the supply line voltage. In the second half of each cycle the switches change over and connect the positive end of C3 to the 0V line and connect the other end of C3 to pin-5, producing a –3V output on pin-5 and thus producing a total of 6V (double the input voltage) between pins 8 and 5. This 6V voltage-doubled output voltage is used to power the IC2 micropower flasher circuit via C2.

The total current drawn from the 3V battery in the *Figure 2.28* circuit equals double that of the 6V micropower flasher (12μA), plus the 'unloaded quiescent' current of IC2 (typically about 11μA when using a 3V supply at normal room temperatures), and usually totals about 35μA.

If you have some rare application where you need to power a micropower LED flasher from a 1.5V battery, you have two basic options. The cheapest option is to power the 3V version of the 7555–based *Figure 2.25* circuit from

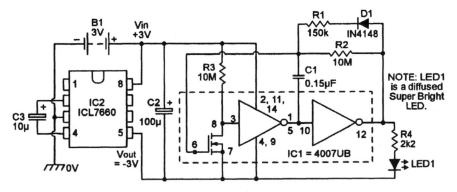

Figure 2.28 *This 4007UB-based micropower LED flasher circuit consumes a mean current of 35μA from a 3V battery*

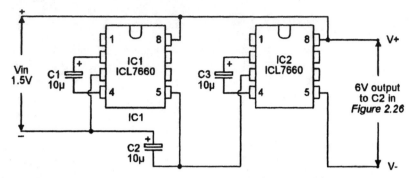

Figure 2.29 *This cascaded ICL7660 voltage doubler gives a 6V output from a 1.5V battery or cell*

the 1.5V source via the basic IC2–plus-C3 voltage doubler used in *Figure 2.28*. In this case the total current drawn from the 1.5V battery is about 142μA.

If you want to power a micropower LED flasher from a 1.5V battery with a minimum of drain current, the best (but rather expensive) option is to power the 6V *Figure 2.26* micropower circuit from the 1.5V battery via a cascaded pair of ICL7660 voltage doubler circuits, connected as shown in *Figure 2.29*. In this case the total current drawn from the 1.5V battery is four-times that of the 6V micropower flasher, plus the 'unloaded quiescent' current of IC2 (typically about 5μA when using a 1.5V supply at normal room temperatures), and totals about 53μA and will give about three years of continuous operation from an 'AA' sized alkaline battery.

3

LED chaser/sequencer circuits

The so-called chaser or sequencer is one of the most popular types of LED-driving circuit and is widely used in advertising displays and in running-light 'rope' displays in small discos, etc. It consists – in essence – of a clocked IC or other electronic unit that drives an array of LEDs in such a way that individual LEDs (or small groups of LEDs) turn on and off in a predetermined and repeating sequence, thus producing a visually attractive display in which one or more ripples of light seem to repeatedly run through a chain or around a ring of LEDs.

The 4017B CMOS IC is probably the best known and most widely used LED-driving IC used in chaser/sequencer applications. This chapter looks at a variety of practical circuits based on this particular IC.

4017B Basics

The 4017B is a member of the popular 4000B family of CMOS digital ICs and can use any DC supply voltage in the 3V to 15V range. It is actually a clocked decade counter/divider IC with ten fully decoded short-circuit-proof outputs that can each be used to drive directly a simple LED display. If desired, various outputs can be coupled back to the IC control terminals to make the device count to (or divide by) any number from 2 to 9 and then either stop or re-start another counting cycle. Numbers of 4017B ICs can be cascaded to give either multi-decade division or to make counters with any desired number of decoded outputs. The 4017B is thus an exceptionally versatile device that can easily be used to chase or sequence a basic LED display of virtually any desired length.

Figure 3.1 shows the outline, pin notations, and basic functional diagram of the 4017B, and *Figure 3.2* shows the waveform timing diagrams of the IC, which incorporates a 5–stage Johnson counter and has CLOCK, RESET and CLOCK INHIBIT input terminals. The internal counters are advanced one

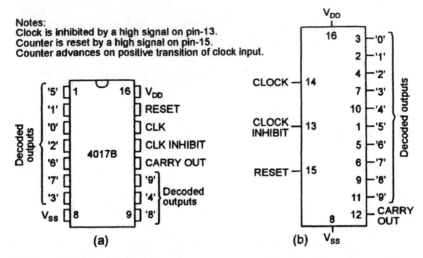

Figure 3.1 *Outline and pin designations (a) and basic functional diagram (b) of the 4017B decade counter/divider IC*

count at each positive transition of the input clock signal when the CLOCK INHIBIT and RESET terminals are low. Nine of the ten decoded outputs are low, with the remaining output high, at any given time. The outputs go high sequentially, in step with the clock signal, with the selected output remaining high for one full clock cycle. An additional CARRY OUT signal completes one cycle for every ten clock input cycles, and can be used to ripple-clock additional 4017B ICs in multi-decade counting applications.

Note that the 4017B counting cycle can be inhibited by setting the CLOCK INHIBIT terminal (pin-13) high, and that a high signal on the RESET terminal (pin-15) clears the counter to zero and sets the decoded '0' output terminal (pin-3) high.

A 4017B LED-driving test circuit

The 4017B is a versatile and easy-to-use IC and (like most 4000B-series ICs) has short-circuit-proof outputs that exhibit slightly surprising characteristics when driving LED-type loads. *Figure 3.3* shows a practical 4017B test circuit that can be used to demonstrate the IC's basic actions and output-driving characteristics. The circuit is best built on a 'plugblock' type of breadboard unit, in which components and wires are simply pushed into the unit's sprung-contact blocks.

In *Figure 3.3*, the 555 timer IC (IC1) is used as a variable-frequency asymmetrical squarewave generator that feeds clocking signals to the CLK input terminal of the 4017B IC (IC2). This output waveform is normally high

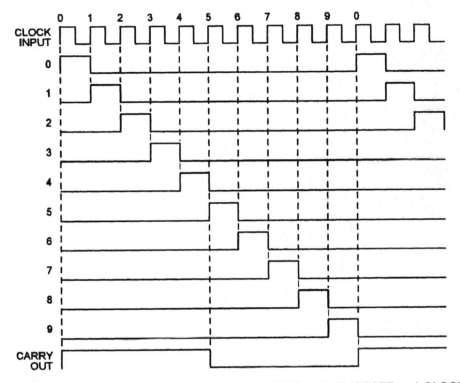

Figure 3.2 *Waveform timing diagram of the 4017B with its RESET and CLOCK INHIBIT terminals grounded*

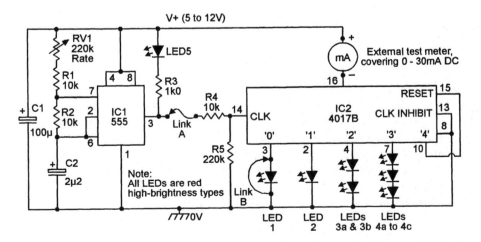

Figure 3.3 *A 4017B LED chaser/sequencer test and demonstration circuit*

but briefly flips low once per cycle and drives LED5 on; the 4017B's internal switching actions are initiated as this signal flips high again and LED5 switches off. Note that the clocking signal is fed to the 4017B IC via removable Link A, and can thus be physically interrupted whenever required; R4 and R5 protect the 4017B's input against damage when Link A is open or IC2's positive supply connection is broken.

In *Figure 3.3*, the positive DC supply line is connected to pin-16 of the 4017B IC via an external multi-range DC current meter that (since IC2's quiescent current is negligible) gives a direct readout of the current drawn by the IC's *currently active* output load. The 4017B is wired (via pins 10 and 15) in the 'divide-by-four' mode and sequentially drives four sets of output loads, which are notated '0' to '3'. Output '0' takes the form of a single LED when Link B is open, or a short-circuit when Link B is closed. Output '1' takes the form of a single LED. Output '2' takes the form of two series-connected LEDs. Output '3' takes the form of three series-connected LEDs. All LEDs are red high-brightness types.

When construction of the *Figure 3.3* circuit is complete, close Link A, open Link B, connect the meter in place, and connect the unit to a 9V DC supply. Adjust RV1 to give a slow clocking rate, noting that LED5 gives a brief flash during each cycle, and that all other LEDs or groups of LEDs activate sequentially. You will probably be surprised to note that all of the display LEDs (LEDs 1 to 4) operate at almost equal brightness, and that all output loads produce fairly similar current readings on the test meter.

When testing the *Figure 3.3* circuit, you can check the individual load currents by waiting until the load activates and then freezing the display by opening Link A. When load '0' is active, the load current is typically 17.5mA with Link B open or 19mA with Link B closed; the load '2' and load '3' currents are typically 16mA and 12.5mA respectively. Thus, when using a 9V supply, the load current is typically 19mA when driving a short-circuit, or 12.5mA when driving three series-connected red LEDs. The graphs of *Figures 3.4* and *3.5* help explain this circuit action.

Figure 3.4 shows the typical forward current/voltage graph of a high-brightness red LED. Note that large variations in forward current produce relatively small variations in forward voltage. Thus, when the current is increased from 10mA to 30mA, the forward voltage increases by only 0.22V, and in this case the LED thus acts like a pure voltage (zero impedance) load in series with an 11 ohm impedance; in practice, this impedance varied between 10 and 15 ohms over most of the LED's working current range.

Figure 3.5 shows the typical supply voltage versus output current graph that applies to each output of the *Figure 3.3* circuit when driving different types of load. Note that each CMOS output stage acts like a loosely controlled constant-current generator that has its short-circuit output current determined by the supply voltage value but has its LED-driving current value influenced by the actual V_{out} value of the stage.

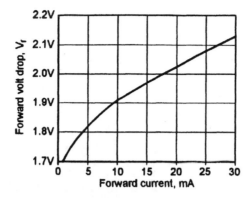

Figure 3.4 *Typical forward current/voltage graph of a high-brightness red LED*

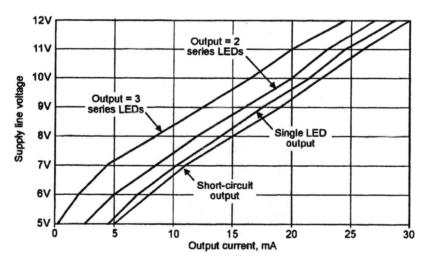

Figure 3.5 *Typical supply voltage versus output current graph of the* Figure 3.3 *circuit when driving different types of load*

In the *Figure 3.3* circuit, when using a 9V supply, V_{out} is zero when driving a shorted output, and under this condition 9V is developed across the output stage, I_{out} is 19mA, and 171mW is thus dissipated in the output stage. When, on the other hand, the 9V circuit is driving three series-connected LEDs, I_{out} is 12.5mA, V_{out} is 5.85V (see *Figure 3.4*), 3.15V is developed across the output stage, and less than 40mW is thus dissipated in the output stage.

Note that most 4000B-series CMOS data sheets list the maximum permitted DC power dissipation values of the 4017B IC as 100mW per-output-stage and 500mW per-package, and these figures should be kept in mind when experimenting with the *Figure 3.3* test/demonstration circuit.

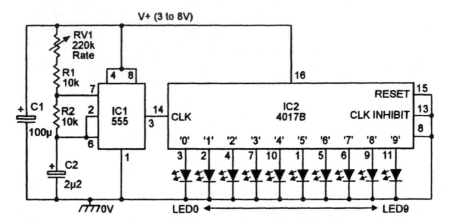

Figure 3.6   *10-LED chaser/sequencer can be used with supply voltages up to only 8V and produces a moving dot display*

Practical 4017B chaser/sequencer circuits

Figure 3.6 shows the practical circuit of a 4017B 10-LED chaser in which IC1 acts as a variable-rate clocking generator and the 4017B IC is wired into the decade counter mode by grounding its CLOCK INHIBIT (pin-13) and RESET (pin-15) control terminals. The circuit action is such that the visual display appears as a moving dot that repeatedly sweeps from the left (LED0) to the right (LED9) in ten discrete steps as the 4017B outputs sequentially go high and drive the LEDs on. The LEDs do not, of course, have to be connected in a straight line; they can, for example, be arranged in a circle, in which case the circle will seem to rotate.

Note that the *Figure 3.6* circuit relies on the internal action of the 4017B to limit the LED currents to safe values, and this circuit can thus be safely used with supply voltages up to a maximum of only 8V without risk of exceeding the IC's 100mW per-output-stage power dissipation limits.

Figure 3.7 shows a modified version of the above circuit, in which a current-limiting 470 ohm resistor is wired in series with each LED to help reduce the IC's power dissipation to a safe level. This circuit can use any DC supply in the 6V to 15V range.

Figure 3.8 shows a circuit variant in which the LEDs share a single current-limiting resistor (R3) and which can be used with reasonable confidence at supply values up to 12V maximum. *Figure 3.9* shows a possible equivalent of this circuit when it is powered from a 15V supply and which illustrates the limitation of the design. The action of the 4017B is such that, when a given LED is on, it effectively grounds the anodes of all other LEDs; R3 thus causes the 'off' LEDs to be reverse biased. Because of the low reverse-voltage ratings of LEDs, this action can cause one or more of the 'off' LEDs

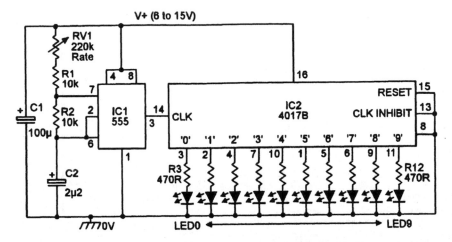

Figure 3.7 *This version of the 10-LED chaser can be used with any supply up to 15V*

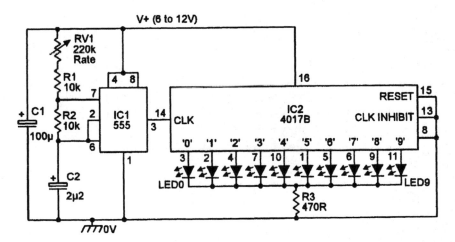

Figure 3.8 *This version of the chaser can be used with supplies up to 12V maximum*

to zener at about 5V, thus giving the results shown in the diagram and possibly causing a power overload in the IC's active output stage.

Thus, when the 4017B is used to drive simple '1-LED-per-output' displays in the moving dot mode, the LEDs can be connected directly to the IC outputs if supply values are limited to 8V maximum, but at supply voltages greater than 8V the LEDs must be connected to the IC outputs via current-limiting resistors. A variety of alternative types of 4017B LED display circuits are shown in *Figures 3.10* to *3.15*.

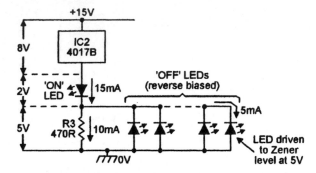

Figure 3.9   *Possible equivalent of the* Figure 3.8 *circuit when powered from a 15V supply*

Alternative LED displays

The output stages of the 4017B can source or sink current with equal ease. *Figure 3.10* shows how the IC can be used in the sink mode to make a moving hole display in which nine of the ten LEDs are on at any given time, with single LEDs turning off sequentially; if the LEDs are wired in the form of a circle, the circle will seem to rotate. Note that, since all LEDs except one are on at the same time, each LED must be provided with a current-limiting resistor to keep the IC power dissipation within safe limits.

In practice, moving dot displays are far more popular than moving hole types. If desired, moving dot displays of the *Figure 3.6* type can be used with fewer than ten LEDs by simply omitting the unwanted LEDs, but in this case the dot will seem to move intermittently, or to scan, since the IC takes ten

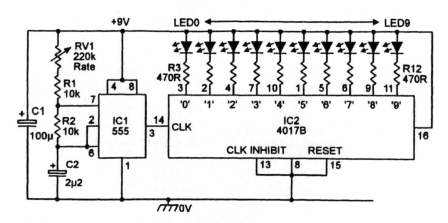

Figure 3.10   *A 10-LED moving hole display*

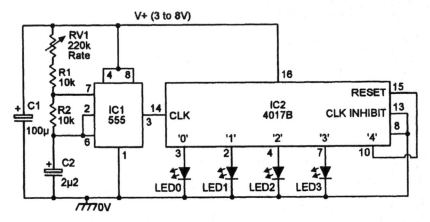

Figure 3.11 *4-LED continuous moving dot display*

clock steps to completely sequence and all LEDs will thus be off during the unwanted steps.

If a continuously moving less-than-10-LED display is wanted it can be obtained by wiring the first unused output terminal of the 4017B to its pin-15 RESET terminal, as shown, for example, in the 4-LED circuit of *Figure 3.11*. Alternatively, the circuit can be made to give an intermittent display with a controlled number of OFF steps by simply taking the appropriate one of the unwanted outputs to the pin-15 RESET terminal. In *Figure 3.12*, for example, the LEDs display for four steps and then blank for four steps, after which the sequence repeats, thus giving a moving dot display with a 50 percent blank period.

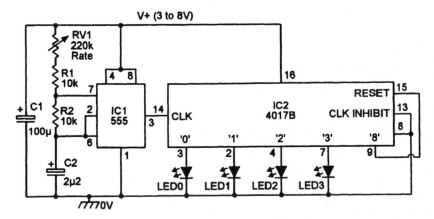

Figure 3.12 *4-LED intermittent moving dot display with 50% blank period*

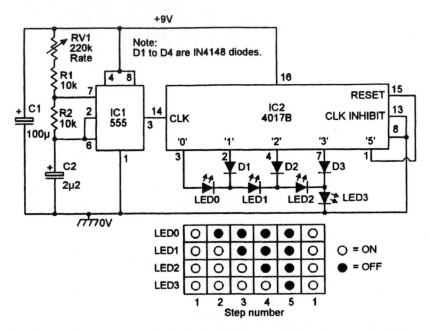

Figure 3.13   *Circuit and performance table of a 4-LED 5–step sequential turn-off display*

Figure 3.13 shows a rather unusual and very attractive 4-LED 5–step sequencer in which all four LEDs are initially on but then turn off one at a time until eventually (in the fifth step) all four LEDs are off; the sequencing details are given in the table of *Figure 3.13*. Note in this circuit that the LEDs are effectively wired in series and that the basic circuit cannot be used to drive more than four LEDs.

Figure 3.14 shows another unusual and attractive LED display. In this case the 4017B runs through a 10–step sequence, with LED1 on for steps 0 to 3, LED2 on for steps 4 to 6, LED3 on for steps 7 and 8, and LED4 on for step 9. The consequence of this action is that the visual display seems to accelerate from LED1 to LED4, rather than sweeping smoothly from one LED to the next. The acceleration action repeats in each switching cycle, and the cycles repeat ad infinitum.

Finally, *Figure 3.15* shows the circuit of a 4–bank 5–step 20-LED chaser that can be used as the basis of a variety of attractive LED displays. Note that a bank of four LEDs are wired in series in each of the five used outputs of the IC, so four LEDs are illuminated at any given time. Roughly 2V are dropped across each ON LED, giving a total drop of 8V across each ON bank, and the circuit's supply voltage must thus be greater than this value for the circuit to operate. A greater number of LEDs can be used in each bank if the supply voltage value is suitably increased.

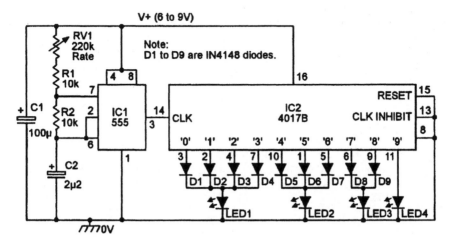

Figure 3.14 *4-LED continuous accelerator display in which the pattern seems to accelerate from left to right*

One of the most attractive and popular LED sequencer displays is the 'light-rope' type, and *Figure 3.16* shows the basic method of constructing a 5-strand 20-LED light-rope display that can be driven by the *Figure 3.15* chaser circuit. Here, each group of four series-connected 'step' output LEDs of the *Figure 3.15* chaser circuit forms one 'strand' of the light rope. There are five strands, and each one must be colour-coded to enable it to be

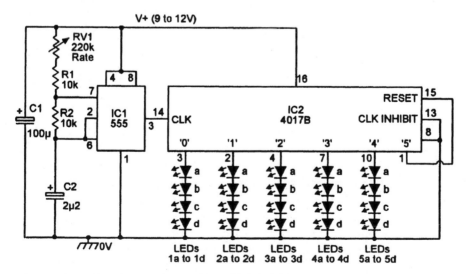

Figure 3.15 *This 4–bank 5-step 20-LED chaser must be used with supply voltages of at least 9V*

connected to the correct output pin of the 4017B IC. In each strand, the four LEDs are evenly spaced apart, but are offset relative to the other four strands, so that there is an equal spacing between all twenty LEDs when the five strands are wrapped together (as shown at the bottom of *Figure 3.16*) to form the complete light-rope, which is usually threaded through a length of protective clear plastic tubing.

If a light-rope of this type uses a fixed spacing of (say) five inches between its LEDs, it will have a total length (allowing for a few unused inches at each end) of about eight feet. When the display is active, four evenly spaced ripples of light seem to run continuously along the length of the light-rope, which is driven directly from the output of the *Figure 3.15* chaser circuit.

Display multiplexing

The basic action of the *Figure 3.14* 4-LED 'accelerator' circuit is such that the light display seems to repeatedly accelerate from left to right, taking a total of ten clock cycles to complete each sequence. *Figure 3.17* shows how the circuit can be modified to give an intermittent display in which the visual acceleration action occurs for ten clock cycles, but all LEDs then blank for the next twenty cycles, after which the action repeats. The circuit action is as follows.

The 4017B has a CARRY OUT terminal on pin-12. When the IC is used in the normal divide-by-10 mode, this CARRY OUT terminal produces one output cycle each time the IC completes a decade count. In *Figure 3.17* this

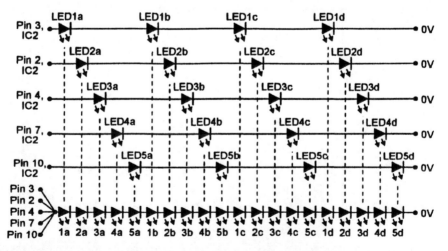

Figure 3.16 *Basic method of constructing a 5–strand 20-LED light-rope display for use with the* Figure 3.15 *circuit*

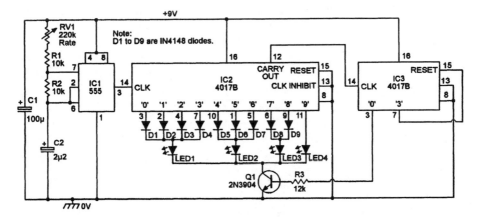

Figure 3.17 *4-LED intermittent accelerator display in which acceleration occurs for ten clock steps in every thirty*

signal is used to clock a second 4017B (IC3), which is wired in the divide-by-3 mode with its '0' output fed to the base of gating transistor Q1. Consequently, during the first ten clock cycles of a sequence the '0' output of IC3 is high and Q1 is biased on, so IC2 acts in the basic manner already described for *Figure 3.14*, with its LEDs turning on sequentially and passing current to ground via Q1. After the tenth clock pulse, however, the '0' output of IC3 goes low and turns Q1 off, so the LEDs no longer illuminate even though IC2 continues to sequence. Eventually, after the 30th clock pulse, the '0' output of IC3 again goes high and turns Q1 on, enabling the display action to repeat again, and so on.

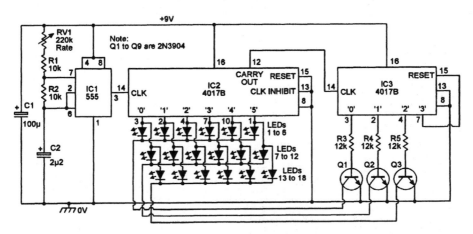

Figure 3.18 *Multiplexed 6-LED × 3-line moving dot display. The dot moves intermittently along the lines*

The *Figure 3.17* circuit is a simple example of display multiplexing, in which IC3 and Q1 are used to selectively enable or disable a bank of LEDs. To conclude this chapter, *Figure 3.18* shows another example of a display multiplexing circuit. In this case the display consists of three lines of six intermittently sequenced LEDs, and these lines are sequentially enabled via IC3 and individual gating transistors, with only one line enabled at any one time.

Note that the basic *Figure 3.18* circuit can easily be expanded to control up to ten sequentially activated lines, which can each have up to ten LED-driving outputs. The expanded circuit can thus be used as a chaser/sequencer with up to 100 LED-driving outputs.

4

LED 'graph' circuits

The previous two chapters took detailed looks at LED principles and at various practical LED visual display circuits. The present chapter continues the LED display theme by looking at a variety of practical LED dot-graph and bar-graph analogue-value display circuits.

LED 'graph' displays

One of the most popular types of multi-LED indicator circuit is the so called analogue-value indicator or 'graph' display, which is designed to drive a chain of linearly spaced LEDs in such a way that the length of chain that is illuminated is proportional to the analogue value of a voltage applied to the input of the LED-driver circuit, e.g. so that the circuit acts like an analogue voltmeter.

Practical graph circuits may be designed to generate either a bar-graph or a dot-graph display. *Figure 4.1* illustrates the bar-graph principle, and shows a line of ten LEDs used to represent a linear-scale 0–10V meter that is indicating (a) 7V or (b) 4V; the input voltage value is indicated by the total number of LEDs that are illuminated. *Figure 4.2* shows the same meter

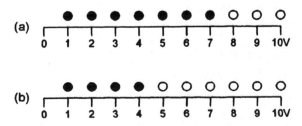

Figure 4.1 *Bar-graph indication of (a) 7V and (b) 4V on a 10V 10-LED scale*

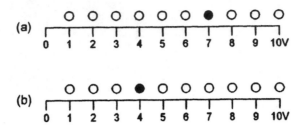

Figure 4.2   *Dot-graph indication of (a) 7V and (b) 4V on a 10V 10-LED scale*

operating in the dot-graph mode; the input voltage value is indicated by the relative position of a single illuminated LED. In reality, the '0' position is often indicated on these scales by a separate LED that is permanently active whenever the display is in use

A number of special ICs are available for operating general-purpose LED analogue-value display systems. For many years the best-known ICs of this type were the U237 (etc.) family from AEG, the UAA170 (etc.) family from Siemens, and the LM3914 (etc.) family from National Semiconductors, but the first two of these families have now ceased production, and only the LM3914 family remains. The LM3914 family are popular and versatile ICs that can each directly drive up to ten LEDs (but can easily be cascaded to drive larger numbers of LEDs) and can drive them in either bar or dot mode.

IC-driven bar-graph displays make inexpensive and in some ways superior alternatives to analogue-indicating moving-coil meters. They are immune to 'sticking' problems, are fast acting, and are unaffected by vibration or by physical attitude. Their scales can easily be given any desired shape. In a given display, individual LED colours can be mixed to emphasize particular sections of the display, and over-range detectors can easily be activated from the driver ICs and used to sound an alarm and/or flash the entire display under the over-range condition.

LED graph displays have better linearity than conventional moving-coil meters, typical linear accuracy being 0.5%. The scale's resolution depends on the number of LEDs used; a 10-LED display gives adequate resolution for many practical purposes. A wide variety of multi-LED LM3914–based graph display circuits are shown in this chapter.

LM3914–family basics

The LM3914 family of dot-/bar-graph driver ICs are manufactured by National Semiconductors. They are moderately complex but highly versatile devices, housed in 18-pin DIL packages and each capable of directly driving up to ten LEDs in either the dot or the bar mode. The family comprises

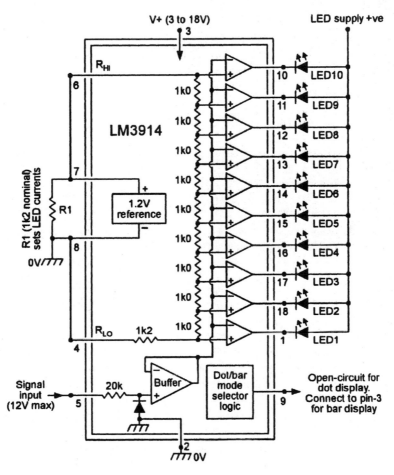

Figure 4.3 *Internal circuit of the LM3914, with connections for making a 10-LED 0–1.2V linear meter with dot- or bar-graph display*

three devices, these being the LM3914, the LM3915, and the LM3916; they all use the same basic internal circuitry (see *Figure 4.3*), but differ in the style of scaling of the LED-driving output circuitry, as shown in *Figure 4.4*.

Thus, the LM3914 is a linearly scaled unit, specifically intended for use in LED voltmeter applications in which the number of illuminated LEDs gives a direct indication of the value of an input voltage (or of some parameter that is represented by a proportional voltage). The LM3915, on the other hand, has a log-scaled output designed to span –27dB to 0dB in ten –3dB steps, and is specifically designed for use in power-indicating applications, etc. Finally, the LM3916 has a semi-log scale that spans 23dB, and is specifically designed for use in VU meter applications.

LED number	Typical threshold-point values, for 10V f.s.d.					
	LM3914 V	LM3915		LM3916		
		V	dB	V	dB	VU
1	1.00	0.447	-27	0.708	-23	-20
2	2.00	0.631	-24	2.239	-13	-10
3	3.00	0.891	-21	3.162	-10	-7
4	4.00	1.259	-18	3.981	-8	-5
5	5.00	1.778	-15	5.012	-6	-3
6	6.00	2.512	-12	6.310	-4	-1
7	7.00	3.548	-9	7.079	-3	0
8	8.00	5.012	-6	7.943	-2	+1
9	9.00	7.079	-3	8.913	-1	+2
10	10.00	10.000	0	10.000	0	+3

Figure 4.4 *Threshold-point values of the LM3914/15/16 range of ICs when designed to drive ten LEDs at a full-scale sensitivity of 10V*

All three devices of the LM3914 family use the same basic internal circuitry, and *Figure 4.3* shows the specific internal circuit of the linear-scaled LM3914, together with the connections for making it act as a simple 10-LED 0–1.2V meter. The IC contains ten voltage comparators, each with its non-inverting terminal taken to a specific tap on a floating precision multi-stage potential divider and with all inverting terminals wired in parallel and accessible via input pin-5 and a built-in unity gain buffer amplifier. The output of each comparator is externally available, and can sink up to 30mA; the sink currents are internally limited, and can be externally pre-set via a single resistor (R1).

The IC also contains a floating 1.2V reference source between pins 7 and 8. In *Figure 4.3* the reference is shown externally connected to the internal potential divider (pins 4 and 6). Note that pins 8 and 4 are shown grounded, so in this case the bottom of the divider is at zero volts and the top is at 1.2V. The IC also contains a logic network that can be externally set (via pin-9) to give either a dot or a bar display from the outputs of the ten comparators. The IC operates as follows.

Assume that the IC logic is set for bar-mode operation, and that the 1.2V reference is applied across the internal 10-stage divider as shown. Thus, 0.12V is applied to the inverting or reference input of the lower comparator, 0.24V to the next, 0.36V to the next, and so on. If a slowly rising input voltage is now applied to pin-5 of the IC the following sequence of actions takes place.

When the input voltage is zero the outputs of all ten comparators are disabled and all LEDs are off. When the input voltage reaches the 0.12V reference value of the first comparator its output conducts and turns LED1 on. When the input reaches the 0.24V reference value of the second comparator its output also conducts and turns on LED2, so at this stage LEDs 1 and 2 are both on. As the input voltage is further increased progressively more and more comparators and LEDs are turned on until eventually, when the input rises to 1.2V, the last comparator and LED10 turn on, at which point all LEDs are on.

A similar kind of action is obtained when the LM3914 logic is set for dot-mode operation, except that only one LED is on at any given time; at zero volts no LEDs are on, and at 1.2V and greater only LED10 is on.

Some finer details

In *Figure 4.3*, R1 is shown connected between pins 7 and 8 (the output of the 1.2V reference), and determines the ON currents of the LEDs. The ON current of each LED is roughly ten times the output current of the 1.2V source, which can supply up to 3mA and thus enables LED currents of up to 30mA to be set via R1. If, for example, a total resistance of 1k2 (equal to the paralleled values of R1 and the 10k of the IC's internal potential divider) is placed across pins 7 and 8 the 1.2V source will pass 1mA and each LED will pass 10mA in the ON mode.

Note from the above that the IC can pass total currents up to 300mA when used in the bar mode with all ten LEDs on. The IC has a maximum power rating of only 660mW, however, so there is a danger of exceeding this rating when the IC is used in the bar mode. In practice, the IC can be powered from DC supplies in the range 3 to 25 volts, and the LEDs can use the same supply as the IC or can be independently powered; this latter option can be used to keep the IC power dissipation at a minimal level.

The internal 10–stage potential divider of the IC is floating, with both ends externally available for maximum versatility, and can be powered from either the internal reference or from an external source or sources. If, for example, the top of the chain is connected to a 10V source, the IC will function as a 0–10V meter if the low end of the chain is grounded, or as a restricted-range 5–10V meter if the low end of the chain is tied to a 5V source. The only constraint on using the divider is that its voltage must not be greater than 2V less than the IC's supply voltage (which is limited to 25V maximum). The input (pin-5) to the IC is fully protected against overload voltages up to plus or minus 35V.

The internal voltage reference of the IC produces a nominal output of 1.28V (limits are 1.2V to 1.34V), but can be externally programmed to produce effective reference values up to 12V (as shown later).

The IC can be made to give a bar display by wiring pin-9 directly to pin-3 (positive supply), or – if only one IC is use – can be made to give a dot display by leaving pin-9 open-circuit or by pulling it at least 200mV below the pin-3 voltage value. If two or more ICs are cascaded to drive 20 or more

LEDs in the dot mode, pin-9 must (except in the case of the final IC in the chain) be wired to pin-1 of the following IC, and a 20k resistor must be wired between pin-11 and the LED-powering positive supply rail.

Finally, note that the major difference between the three members of the LM3914 family of ICs lays in the values of resistance used in the internal 10–stage potential divider. In the LM3914 all resistors in the chain have equal values, and thus produce a linear display of ten equal steps. In the LM3915 the resistors are logarithmically weighted, and thus produce a log display that spans –27dB to 0dB in ten –3dB steps. In the LM3916 the resistors are weighted in semi-log fashion and produce a display that is specifically suited to VU-meter applications. Let's now move on and look at some practical applications of this series of devices, paying particular attention to the linear LM3914 IC.

Dot-mode voltmeters

Figures 4.5 to *4.9* show various ways of using the LM3914 IC to make 10-LED dot-mode voltmeters with a variety of full-scale deflection (f.s.d.) sensitivities. Note in all these circuits that pin-9 is left open-circuit, to give dot-mode operation, and that a 10μF capacitor is wired directly between pins 2 and 3 to enhance circuit stability.

Figure 4.5 shows the connections for making a variable-range (1.2V to 1000V f.s.d.) voltmeter. The low ends of the internal reference and divider are grounded and their top ends are joined together, so the meter has a basic full-scale sensitivity of 1.2V, but variable ranging is provided by the Rx–R1 potential divider at the input of the circuit. Thus, when Rx is zero, f.s.d. is

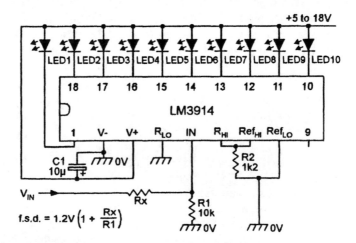

Figure 4.5   *1.2V to 1000V f.s.d. dot-mode voltmeter*

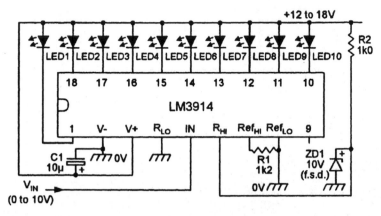

Figure 4.6 *10V f.s.d. meter using an external reference*

1.2V, but when Rx is 90k the f.s.d. is 12V. Resistor R2 is wired across the internal reference and sets the ON currents of the LEDs at about 10mA.

Figure 4.6 shows how to make a fixed-range 0–10V meter, using an external 10V zener (connected to the top of the internal divider) to provide a reference voltage. The supply voltage to this circuit must be at least two volts greater than the zener reference voltage.

Figure 4.7 shows how the internal reference of the IC can be made to effectively provide a variable voltage, enabling the meter f.s.d. value to be set anywhere in the range 1.2V to 10V. In this case the 1mA current (determined by R1) of the floating 1.2V internal reference flows to ground via RV1, and the resulting RV1 voltage raises the reference pins (pins 7 and 8) above zero. If, for example, RV1 is set to 2k4, pin 8 will be at 2.4V and pin-7 at 3.6V. RV1 thus enables the pin-7 voltage (connected to the top of the

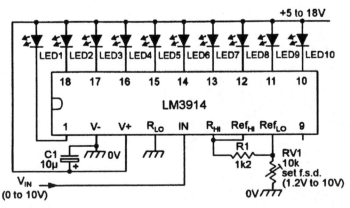

Figure 4.7 *An alternative variable-range (1.2V to 10V) dot-mode voltmeter*

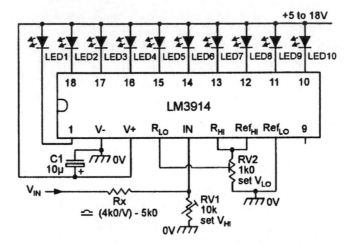

Figure 4.8 *Expanded-scale (10V – 15V, etc.) dot-mode voltmeter*

internal divider) to be varied from 1.2V to about 10V, and thus sets the f.s.d. value of the meter within these values. Note that the circuit's supply voltage must be at least 2V greater than the desired f.s.d. voltage value.

Figure 4.8 shows the connections for making an expanded-scale meter that, for example, reads voltages in the range 10 to 15 volts. RV2 sets the LED current at about 12mA, but also enables a reference value in the range 0–1.2V

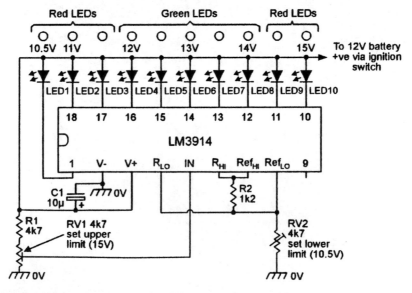

Figure 4.9 *Expanded-scale dot-mode vehicle voltmeter*

to be set on the low (pin-4) end of the internal divider. Thus, if RV2 is set to apply 0.8V to pin-4 the basic meter will read voltages in the range 0.8 to 1.2 volts only. By fitting potential divider Rx–RV1 to the input of the circuit this range can be amplified to (say) 10–15V, or whatever range is desired.

Finally, *Figure 4.9* shows an expanded scale dot-mode voltmeter that is specifically designed to indicate the value of a vehicle's battery (12V nominal). In this case R2–RV2 are effectively set to give a basic range of 2.4 to 3.6 volts, but the input to the circuit is derived from the positive supply rail via the R1–RV1 potential divider, and the indicated volts reading thus corresponds to a pre-set multiple of the basic range value. As shown in the diagram, red and green LEDs can be used in the display, arranged so that green LEDs illuminate when the voltage is in the safe range 12 to 14 volts.

To calibrate the above circuit, first set the supply to 15V and adjust RV1 so that LED-10 just turns on. Reduce the supply to 10V and adjust RV2 so that LED-1 just turns on. Recheck the settings of RV1 and RV2. The calibration is then complete and the unit can be installed in the vehicle by taking the 0V lead to chassis and the +12V lead to the vehicle's battery via the ignition switch.

Bar-mode voltmeters

The dot-mode circuits of *Figures 4.5* to *4.9* can be made to give bar-mode operation by simply connecting pin 9 to pin-3, rather than to pin-11. When using the bar mode, however, it must be remembered that the IC's power rating must not be exceeded by allowing excessive output-terminal voltages to be developed when all ten LEDs are on. LEDs drop roughly 2V when

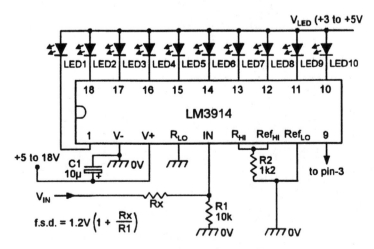

Figure 4.10 *Bar-display voltmeter with separate LED supply*

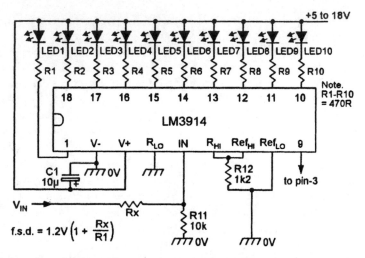

Figure 4.11   *Bar-display voltmeter with common LED/IC supply*

they are conducting, so one way around this problem is to power the LEDs from their own low-voltage (3 to 5V) supply, as shown in *Figure 4.10*.

An alternative solution is to power the IC and the LEDs from the same supply, but to wire a current-limiting resistor in series with each LED, as shown in *Figure 4.11*, so that the IC's output terminal saturates when the LEDs are on.

Figure 4.12 shows another way of obtaining a bar display without excessive power dissipation. Here, the LEDs are all wired in series, but with each

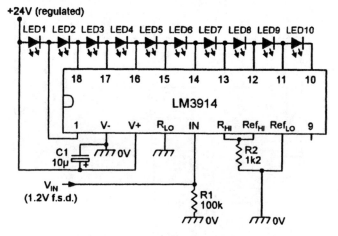

Figure 4.12   *Method of obtaining a bar display with dot-mode operation and minimal current consumption*

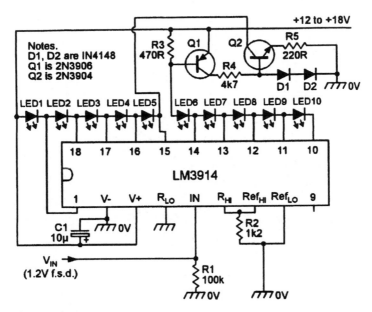

Figure 4.13 *Modification of the Figure 4.12 circuit, for operation from unregulated 12V to 18V supplies*

one connected to an individual output of the IC, and the IC is wired for dot-mode operation. Thus, when (for example) LED-5 is on it draws its current via LEDs 1 to 4, so all five LEDs are on and the total LED current equals that of a single LED, and total power dissipation is quite low. The LED supply to this circuit must be greater than the sum of all LED volt-drops when all LEDs are on, but must be within the voltage limits of the IC; a regulated 24V supply is thus needed.

Figure 4.13 shows a very useful modification which enables the above circuit to be powered from unregulated supplies within the 12 to 18V range. In this case the LEDs are split into two chains, and the transistors are used to switch on the lower (LEDs 1 to 5) chain when the upper chain is active; the maximum total LED current equals twice the current of a single LED.

20-LED voltmeters

Figure 4.14 shows how two LM3914 ICs can be interconnected to make a 20-LED dot-mode voltmeter. Here, the input terminals of the two ICs are wired in parallel, but IC1 is configured so that it reads 0 to 1.2 volts, and IC2 is configured so that it reads 1.2 to 2.4 volts. In the latter case the low end of the IC2 potential divider is coupled to the 1.2V reference of IC1, and the

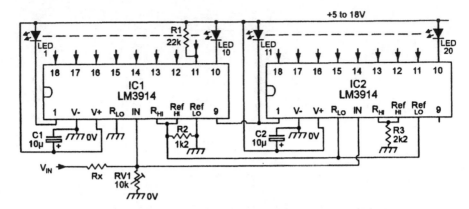

Figure 4.14   *Dot-mode 20-LED voltmeter (f.s.d. = 2.4V when Rx = 0)*

top end of the divider is taken to the top of the 1.2V reference of IC2, which is raised 1.2V above that of IC1.

The 20-LED *Figure 4.14* circuit is wired for dot-mode operation, and in this case pin-9 of IC1 is wired to pin-1 of IC2, pin-9 of IC2 is open-circuit, and a 22k resistor is wired in parallel with LED-9 of IC1.

Figure 4.15 shows the connections for making a 20-LED bar-mode voltmeter. The connections are similar to those of *Figure 4.14*, except that pin-9 is taken to pin-3 on each IC, and a 470R current-limiting resistor is wired in series with each LED to reduce the power dissipation of the ICs.

To conclude this look at LM3914 circuits, *Figure 4.16* shows a simple frequency-to-voltage converter that can be used to convert either of the

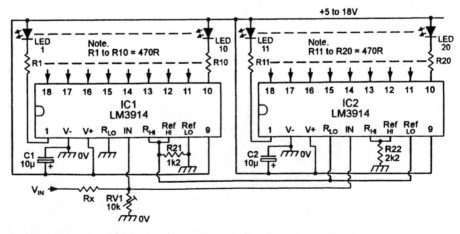

Figure 4.15   *Bar-mode 20-LED voltmeter (f.s.d. = 2.4V when Rx = 0)*

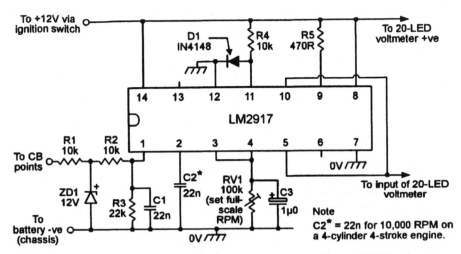

Figure 4.16 *Vehicle tacho conversion circuit for use with a 20-LED voltmeter*

Figure 4.14 or *4.15* circuits into 20-LED tachometers (RPM-meters). This converter should be interposed between the vehicle's contact-breaker points and the input of the voltmeter circuit. *In Figure 4.16*, the C2 value of 22n is the optimum value for a full-scale range of 10 000 RPM on a 4–cylinder 4–stroke engine. For substantially lower full-scale RPM values, the C2 value may have to be increased: the value may have to be reduced on vehicles with six or more cylinders.

LM3915/LM3916 circuits

The LM3915 'log' and LM3916 'semi-log' ICs operate in the same basic way as the LM3914, and can in fact be directly used in most of the circuits shown in *Figures 4.5* to *4.15*. In most practical applications, however, these particular ICs are used to indicate the value of an a.c. input signal, and the simplest way of achieving such a display is to connect the a.c. signal directly or via an attenuator to the pin-5 input terminal of the IC, as shown in *Figure 4.17*. The IC responds only to the positive halves of such input signals, and the number of illuminated LEDs is thus proportional to the instantaneous peak value of the input signal.

The *Figure 4.17* circuit is that of a simple LM3915–based audio power meter that is used to indicate instantaneous output voltage values from an external loudspeaker. Pin-9 is left open-circuit to give dot-mode operation, and R1 has a value of 390R to give a LED current of about 30mA, thus giving a clear indication of brief instantaneous voltage levels. The meter gives audio power indication over the range 200mW to 100W.

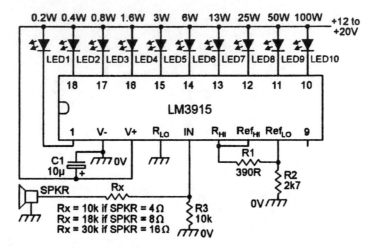

Figure 4.17   *Simple speaker-driven audio power meter*

Figure 4.18 shows the basic way of using the LM3916 IC as a VU-meter with a full-scale sensitivity of 10V d.c. The circuit is shown connected for bar-mode operation, using separate supply voltages for the LED display and for the actual IC, and with the component values shown gives a current drive of 10mA to each active LED. If preferred, the IC can be used to give dot-mode operation, using a common 12V to 20V supply for the LEDs and the IC, by leaving pin-3 open circuit and changing the R1–R2 values to 390R–2k4, thus giving 30mA of drive to the active LEDs.

Figure 4.19 shows an alternative way of using the LM3916 as a VU-meter with a bar-type display. In this case the IC is used in the same way as the

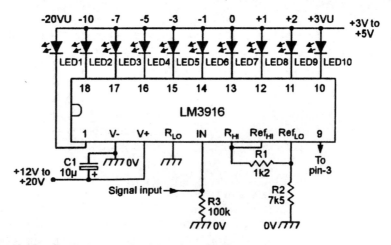

Figure 4.18   *Basic bar-mode VU-meter circuit*

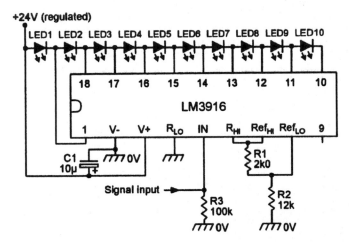

Figure 4.19 *This basic VU-meter circuit gives a bar-type display, with a dot-type current drain*

basic *Figure 4.12* low-current-consumption circuit, with pin-9 left open-circuit so that the IC actually operates in the dot mode, but with the LEDs wired in series across the display-driving pins so that a bar-type display is obtained, with all active LED currents flowing through the currently active driving pin. With the component values shown, this circuit has a full-scale sensitivity of 10V and provides a LED-drive current of 16mA.

The basic *Figure 4.17* to *4.19* LM3915 and LM3916 circuits are shown being driven directly from a.c. signal inputs, and this technique is adequate in many applications. In cases where the display is required to relate specifically to peak, r.m.s., or average values of a.c. input voltage, this can be achieved by interposing a suitable a.c.-to-d.c. converter circuit between the a.c. signal and the pin-5 input terminal of the LM3915 or LM3916 IC. Many suitable circuits are published in op-amp application manuals, circuit reference books and encyclopaedias, etc.

An over-range alarm-driver circuit

To conclude this chapter, *Figure 4.20* shows a simple way of fitting an alarm-driving over-range switch to a bar-type LM3914–series LED-driving indicator circuit. Here, *pnp* transistor Q1 is wired between the LED positive supply rail and the 0V rail, with its base connected to the IC's pin-10 (which drives LED10) and with a self-contained alarm unit wired in series with its collector. Normally, LED10, Q1 and the alarm are all off, but if LED10 turns on it pulls Q1 on via R2 and thus activates the alarm unit, which indicates the over-range condition.

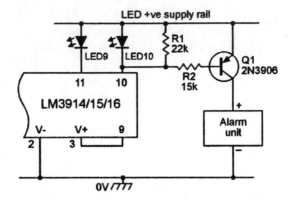

Figure 4.20 *An over-range alarm-driver circuit, for use with bar-type displays*

In this circuit, the alarm unit may take the form of a piezo siren unit that generates an acoustic alarm sound, or a gated astable switch unit that repeatedly switches the LED brightness between high and low levels under the over-range condition, or maybe a combination of both of these units. If desired, the unit can be activated by any one of the display LEDs, in which case the alarm will activate whenever that or any higher LED is energized.

5

7-segment displays

The previous three chapters have described various optoelectronic display circuits based on simple LEDs. The present chapter continues the display theme by looking at 7-segment alphanumeric displays and at a variety of associated display-driver devices and circuitry.

7-segment displays

A very common requirement in modern electronics is that of displaying alphanumeric characters. Digital watches, pocket calculators, and digital multimeters and frequency meters are all examples of devices that use such displays. The best-known type of alphanumeric indicator is the 7-segment display, which comprises seven independently accessible photoelectric segments (such as LEDs or liquid crystals, or gas-discharge or fluorescent elements, etc.) arranged in the form shown in *Figure 5.1*. The segments are conventionally notated from *a* to *g* in the manner shown in the diagram, and it is possible to make them display any number (numeral) from 0 to 9 or alphabetic character from A to F (in a mixture of upper and lower case letters) by activating these segments in various combinations, as shown in the truth table of *Figure 5.2*.

Practical 7-segment display devices must be provided with at least eight external connection terminals; seven of these give access to the individual

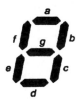

Figure 5.1 *Standard form and notations of a 7-segment display*

Segments (✓ = ON)							Display
a	b	c	d	e	f	g	
✓	✓	✓	✓	✓	✓		0
	✓	✓					1
✓	✓		✓	✓		✓	2
✓	✓	✓	✓			✓	3
	✓	✓			✓	✓	4
✓		✓	✓		✓	✓	5
✓		✓	✓	✓	✓	✓	6
✓	✓	✓					7

Segments (✓ = ON)							Display
a	b	c	d	e	f	g	
✓	✓	✓	✓	✓	✓	✓	8
✓	✓	✓			✓	✓	9
✓	✓	✓		✓	✓	✓	A
		✓	✓	✓	✓	✓	b
✓			✓	✓	✓		C
	✓	✓	✓	✓		✓	d
✓			✓	✓	✓	✓	E
✓				✓	✓	✓	F

Figure 5.2   *Truth table for a 7-segment display*

photoelectric segments, and the eighth provides a common connection to all segments. If the display is a LED type, the seven individual LEDs may be arranged in the form shown in *Figure 5.3*, in which all LED anodes are connected to the common terminal, or they may be arranged as in *Figure 5.4*, in which all LED cathodes are connected to the common terminal. In the former case the device is known as a common-anode 7-segment display; in the latter case the device is known as a common-cathode 7-segment display.

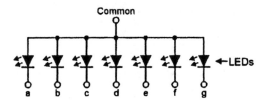

Figure 5.3   *Schematic diagram of a common-anode 7-segment LED display*

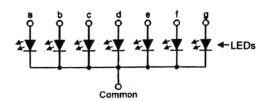

Figure 5.4   *Schematic diagram of a common-cathode 7-segment LED display*

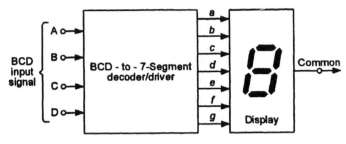

Figure 5.5 *Basic connections of a BCD-to-7-segment decoder/driver IC*

7-segment display/drivers

In most practical applications, 7-segment displays are used to give a visual indication of the output states of digital ICs such as decade counters and latches, etc. These outputs are usually in 4–bit BCD (binary coded decimal) form and are not suitable for directly driving 7-segment displays. Consequently, special BCD-to-7-segment decoder/driver ICs are available to convert the BCD signal into a form suitable for driving these displays, and are connected between the BCD signals and the display in the manner shown in *Figure 5.5*. The table of *Figure 5.6* shows the relationship between the BCD signals and the displayed 7-segment numerals.

In practice, BCD-to-7-segment decoder/driver ICs are usually available in a dedicated form that is suitable for driving only a single class of display unit, e.g. either common-anode LED type, or common-cathode LED type, or liquid crystal displays (LCDs). *Figures 5.7* to *5.9* show the methods of interconnecting each of these IC and display types.

Note in the case of the LED circuits (*Figures 5.7* and *5.8*) that if the IC outputs are unprotected (as in the case of most TTL ICs), a current-limiting

BCD Signal				Display	BCD Signal				Display
D	C	B	A		D	C	B	A	
0	0	0	0	*0*	0	1	0	1	*5*
0	0	0	1	*1*	0	1	1	0	*6*
0	0	1	0	*2*	0	1	1	1	*7*
0	0	1	1	*3*	1	0	0	0	*8*
0	1	0	0	*4*	1	0	0	1	*9*

0 = logic low   1 = logic high

Figure 5.6 *Truth table of a BCD-to-7-segment decoder/driver*

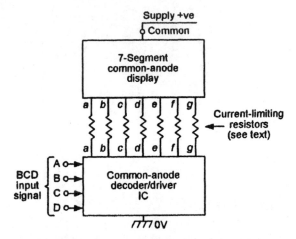

Figure 5.7   *Method of driving a common-anode LED display*

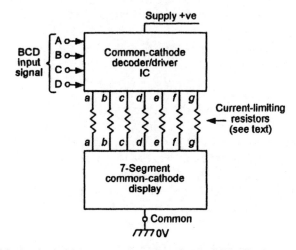

Figure 5.8   *Method of driving a common-cathode LED display*

resistor must be wired in series with each display segment (about 150R with a 5V supply, or 680R at 15V); most CMOS ICs have internally current-limited outputs, and do not require the use of these external resistors. To drive a common-anode display (*Figure 5.7*), the driver must have an active-low output, in which each segment-driving output is normally high, but goes low to turn a segment on. To drive a common-cathode display (*Figure 5.8*), the driver must have an active-high output.

In the *Figure 5.9* LCD-driving circuit the display's common BP (back-plane) terminal and the IC's phase input terminals must be driven by a

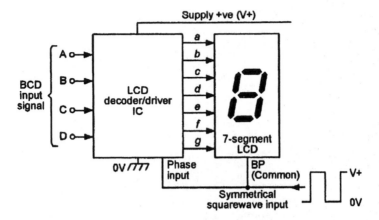

Figure 5.9 *Method of driving a liquid crystal display (LCD)*

symmetrical squarewave (typically 30Hz to 200Hz) that switches fully between the two supply rail voltages (0V and V+), as shown. The full explanation for this is a little complicated, as follows.

To drive a LCD segment, the driving voltage must be applied between the segment and BP terminals. When the voltage is zero, the segment is effectively invisible. When the drive voltage has a significant positive or negative value, however, the segment becomes effectively visible, but if the drive voltage is sustained for more than a few hundred milliseconds the segment may become permanently visible and be of no further value. The way around this problem is, in principle, to drive the segment on via a perfectly symmetrical squarewave that switches alternately between identical positive and negative voltages, and thus has zero dc components and will not damage the LCD segment even if sustained permanently. In practice, this type of waveform is actually generated with the aid of an EX-OR True/Complement generator, connected as shown in *Figure 5.10*(a).

In *Figure 5.10(a)*, the basic segment 'a' input drive (which is active-high) is connected to one input of the EX-OR element, and the other EX-OR input terminal (which is notated PHASE) is driven by a symmetrical squarewave that switches fully between the circuit's supply rail voltages (shown as 0V and +10V) and is also applied to the LCD display's BP pin. When the segment *a* input drive is low, the EX-OR element gives a non-inverted (in-phase) *a* output when the squarewave is at logic-0, and an inverted (anti-phase) *a* output when the squarewave is at logic-1, and thus produces zero voltage difference between the *a* segment and BP points under both these conditions; the segment is thus turned off under these conditions. When the segment *a* input drive is high, the EX-OR element gives the same phase action as just described, but in this case the *a* OUT pin is high and BP is low when the squarewave is at logic-0, and *a* OUT is low and BP is high when

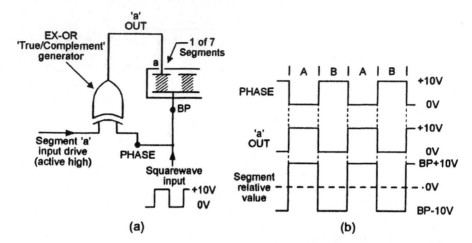

Figure 5.10   *Basic LCD segment-drive circuit (a), and voltage-doubling 'bridge-driven' segment waveforms (b)*

the squarewave is at logic-1; the segment is thus turned on under these conditions.

Figure 5.10(b) shows the circuit waveforms that occur when the *a* segment is turned on, with the *a* segment and BP driven by anti-phase squarewaves. Thus, in part A of the waveform the segment is 10V positive to BP, and in part B it is 10V negative to BP, so the LCD is effectively driven by a squarewave with a peak-to-peak value of 20V but with zero dc value. This form of drive is generally known as a voltage-doubling 'bridge-drive' system. In practice, many LCD-driving ICs (such as the 4543B) incorporate this type of drive system in the form of a 7–section EX-OR gate array interposed in series with the segment output pins, with access to its common line via a single PHASE terminal.

Note that any active-high 7-segment LED-driving decoder IC can be used to drive a 7-segment LCD display by interposing a bridge-driven 7–section EX-OR array between its segment output pins and the segment pins of the LCD display, as shown in *Figure 5.11*, in which a 74LS48 TTL IC is used in this specific way.

Cascaded displays

In most practical 7-segment display applications, several sets of displays and matching decoder/driver ICs are cascaded and used to make multi-digit display systems. *Figure 5.12*, for example, shows a very simple way of using three sets of decoder/driver ICs and displays in conjunction with three decade counter ICs to make a simple digital-readout frequency meter. Here,

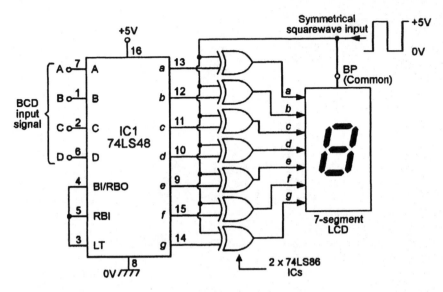

Figure 5.11 *Basic way of using an active-high LED-driving decoder IC (such as the 74LS48 TTL type) to drive a 7-segment LCD via a bridge-driven 7–section EX-OR array*

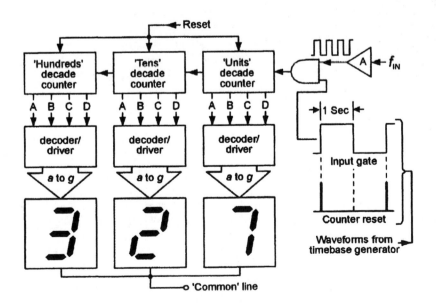

Figure 5.12 *Simple digital frequency meter circuit*

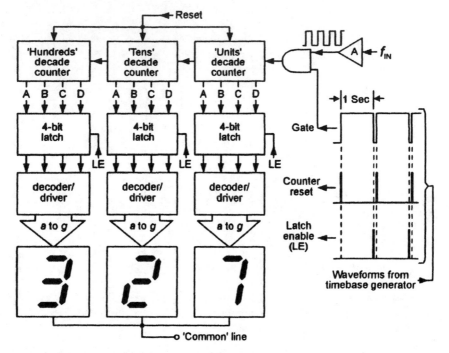

Figure 5.13   *Improved digital frequency meter circuit*

the amplified external frequency signal is fed to the input of the series-connected counters via one input of a 2-input AND gate, which has its other (GATE) input waveform derived from a built-in timebase generator. The circuit's operating sequence is as follows:

When the timebase GATE input signal is low the AND gate is closed and no input signals reach the counters. At the moment that the timebase GATE signal switches high a brief RESET pulse is fed to all three counters, setting them all to zero count; simultaneously, the input gate opens, and remains open for a period of precisely one second, during which time the input-frequency pulses are summed by the counters. At the end of the one second period the gate closes and the timebase GATE signal goes low again, thus ending the count and enabling the displays to give a steady reading of that second's total pulse count (and thus the mean signal frequency). The whole process then repeats again one second later, when the timebase GATE signal again goes high.

Display latching

The simple cascaded system described above suffers from a major defect, in that the display becomes a blur during the actual counting period, becoming stable and readable only when each count is completed and the input gate is closed. This 'blur-and-read' type of display is very annoying to watch. *Figure 5.13* shows an improved frequency meter circuit that uses display latching to overcome the above defect. Here, a 4-bit data latch is wired between the output of each counter and the input of its decoder/driver IC. This circuit operates as follows:

At the moment that the timebase GATE signal goes high a RESET pulse is fed to all counters, setting them to zero. Simultaneously, the input gate is opened and the counters start to sum the input signal pulses. This count continues for precisely one second, and during this period the 4-bit latches prevent the counter output signals from reaching the display drivers; the display thus remains stable during this period.

At the end of the one second count period the AND gate closes and terminates the count, and simultaneously a brief LATCH ENABLE pulse is fed to all latches, causing the prevailing BCD outputs of each counter to be latched into memory and thence fed to the display via the decoder/driver ICs, thus causing the display to give a steady reading of the total pulse count (and thus the input frequency). A few moments later the sequence repeats again, with the counters resetting and then counting the input frequency pulses for one second, during which time the display gives a steady reading of the results of the previous count, and so on.

The *Figure 5.13* circuit thus generates a stable display that is updated once every second; in practice, the actual count period of this and the *Figure 5.12* circuit can be made any decade multiple or submultiple of one second, provided that the output display is suitably scaled. Note that a 3-digit frequency meter can indicate maximum frequencies of 999Hz when using a 1-second timebase, 9.99kHz when using a 100ms timebase, 99.9kHz when using a 10ms timebase, and 999kHz when using a 1ms timebase. Also note that in reality many decoder/driver ICs have built-in 4-bit data latches.

Multiplexing

Note from the *Figure 5.12* and *5.13* circuits that a total of at least 21 connections must be made between the IC circuitry and the 7-segment displays of a 3-digit readout unit; a total of at least 70 connections are needed if a 10-digit display is used. In reality, the number of IC-to-display connections can be greatly reduced by using the technique known as multiplexing. This technique can be understood with the aid of *Figures 5.14* and *5.15*.

Figure 5.14 shows how each digit of a 3-digit common-cathode LED display can be individually activated using a total of only ten external connections; the circuitry to the left of the dotted line should be regarded as 'electronic', and to the right of the line as 'display' circuitry. In the display, all *a* segments are connected together, as also are all other (*b* to *g*) sets of

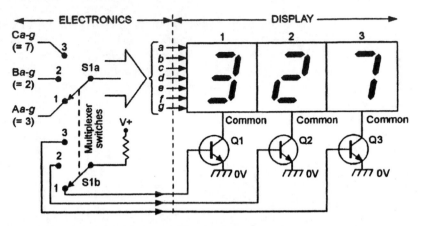

Figure 5.14   *Method of multiplexing a 3–digit common-cathode LED display*

segments, so that a total of only seven external *a* to *g* connections are made to the display irrespective of the number of digits used. Note, however, that none of the 7-segment displays are influenced by signals on these segment wires unless a display is enabled by connecting its common terminal to ground, and in *Figure 5.14* this is achieved by activating switching transistors Q1 to Q3 via suitable external signals, which require the use of only one additional connection per display digit.

Note in *Figure 5.14* that three different sets of segment data can be selected via switch S1a, which in reality would take the form of a ganged 7-pole 3-way electronic switch (with one pole dedicated to each of the seven segment lines), and that any one of the three display digits can be selected via S1b and Q1 to Q3. These switches are ganged together and provide the actual multiplexer action, and should be regarded as fast-acting electronic switches that repeatedly switch through positions 1, 2 and 3. The operating sequence of the circuit is as follows.

Assume initially that the switch is in position 1. Under this condition S1a selects segment data A*a*–*g*, and S1b activates display 1 via Q1, so that display 1 shows the number 3. A few moments later the switch jumps to position 2, selecting segment data B*a*–*g* and activating display 2 via Q2, so that display 2 shows the number 2. A few moments later the switch jumps to position 3, causing display 3 to show the number 7. A few moments later the whole cycle starts to repeat again, and so on ad infinitum. In practice, about fifty of these cycles occur each second, so the eye does not see the displays being turned on and off individually but sees them as an apparently steady display that shows the number 327, or whatever other number is dictated by the segment data.

Note from the above description that, since each display is turned on for only one-third of each cycle, the mean current consumption of each display

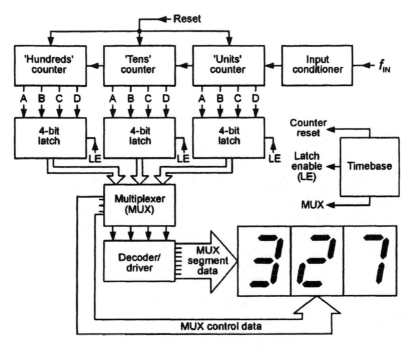

Figure 5.15 *Realistic implementation of the multiplexing technique in a 3–digit frequency meter*

is one-third of the peak display current, and the LED brightness levels are correspondingly reduced. In practical multiplexers the peak display current is made fairly high, to give adequate display brightness.

Figure 5.15 shows an example of an improved multiplexing (MUX) technique, as applied to a 3-digit frequency meter. In this case the MUX is interposed between the outputs of the three BCD data latches and the input of the BCD-to-7-segment decoder/driver IC. This technique has two major advantages. First, it calls for the use of only a single decoder/driver IC, irrespective of the number of readout digits used. Second, it calls for the use of a MUX incorporating only five ganged 3-way sequencing switches (one for the control data and four for the BCD data), rather than the eight ganged 3-way switches (one for the control data and seven for the segment data) called for in the *Figure 5.14* system.

In practice, all of the counting, latching, multiplexing, decoding, timing and display-driving circuitry of *Figure 5.15* (and a great deal more) can easily be incorporated in a single LSI (large scale integration) chip that needs only twenty or so pins to make all necessary connections to the power supply, displays, and inputs, etc. Thus, a complete 4-digit counter can be implemented using a dedicated IC in a circuit such as that shown in *Figure 5.16*,

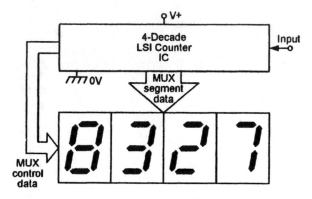

Figure 5.16 *4–digit counter circuit, using a LSI chip*

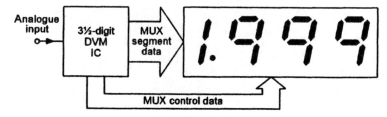

Figure 5.17 *3½-digit DVM using a LSI chip*

or a 3½-digit DVM (digital voltmeter) can be implemented using a circuit such as that shown in *Figure 5.17*.

Ripple blanking

If the basic 4-digit *Figure 5.16* circuit is used to measure a count of 27 it will actually give a reading of 0027, unless steps are taken to provide automatic suppression of the two (unwanted) leading zeros. Similarly, if the 3½-digit circuit of *Figure 5.17* is used to measure 0.1 volts it will actually give a display of 0.100 volts, unless steps are taken to provide automatic suppression of the two (unwanted) trailing zeros.

In practice, automatic blanking of leading and/or trailing zeros can be obtained by using a ripple blanking technique, as illustrated in *Figures 5.18* and *5.19*. In these diagrams, each decoder/driver IC has a BCD input and a 7-segment output, and is provided with ripple blanking input (RBI) and output (RBO) terminals. If these terminals are active high they will have the following characteristics.

If the RBI terminal is held low (at logic-0), the 7-segment outputs of the IC are enabled but the RBO terminal is disabled (held low). If the RBI

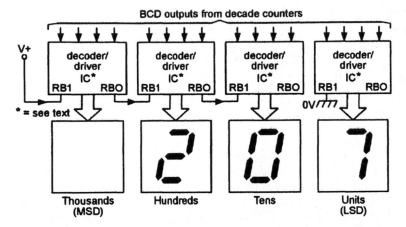

Figure 5.18 *Ripple-blanking used to give leading-zero suppression in a 4–digit counter*

terminal is biased high (at logic-1), the 7-segment outputs become disabled in the presence of a BCD '0000' input (= decimal zero), and the RBO output goes high under the same condition. Thus, the RBO terminal is normally low and goes high only if a BCD '0000' input is present at the same time as the RBI terminal is high. With these characteristics in mind, refer now to *Figures 5.18* and *5.19.*

Figure 5.18 shows the ripple blanking technique used to provide leading-zero suppression in a 4-digit display that is reading a count of 207. Here, the RBI input of the thousands or most significant digit (MSD) decoder/driver is tied

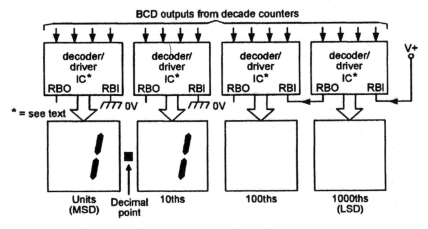

Figure 5.19 *Ripple-blanking used to give trailing-zero suppression of the last two digits of a 3½-digit DVM readout*

high, so this display is automatically blanked in the presence of a zero, under which condition the RBO terminal is high. Consequently, the RBI terminal of the hundreds IC is high, so its display reads 2, and the RBO terminal is low. The RBI input of the tens unit is thus also low, so its display reads 0 and its RBO output is low. The least significant digit (LSD) is that of the units readout, and this does not require zero suppression; consequently, its RBI input is grounded and it reads 7. The display thus gives a total reading of 207.

Note in the *Figure 5.18* leading zero suppression circuit that ripple blanking feedback is applied backwards, from the MSD to the LSD. *Figure 5.19* shows how trailing zero suppression can be obtained by reversing the direction of feedback, from the LSD to the MSD. Thus, when an input of 1.1 volts is fed to this circuit the LSD is blanked, since its BCD input is '0000' and its RBI input is high. Its RBO terminal is high under this condition, so the 100ths digit is also blanked in the presence of a '0000' BCD input.

Practical decoder/driver ICs are often (but not always) provided with ripple blanking input and output terminals; often, these are active low. If a decoder/driver IC does not incorporate integral ripple blanking logic, it can usually be obtained by adding external logic similar to that shown in *Figure 5.20*, with the RBO terminal connected to the BLANKING input pin of the decoder/driver IC. In *Figure 5.20* (an active-high circuit), the output of the 4-input NOR gate goes high only in the presence of a '0000' BCD input, and the RBO output goes high only if the decimal zero input is present while RBI is high.

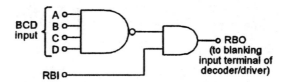

Figure 5.20   *DIY ripple-blanking logic (active-high type)*

Practical decoder/driver ICs

Decoder/driver ICs are available in both TTL and CMOS forms. Some of these devices have integral ripple blanking facilities, others have built-in data latches, and a few even have built-in decade counter stages, etc. The rest of this chapter describes a few of the most popular of these devices.

The 74LS47 and 74LS48.

These 7-segment decoder/driver ICs are members of the LS TTL family. They have integral ripple-blanking facilities, but do not incorporate data

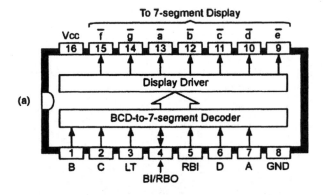

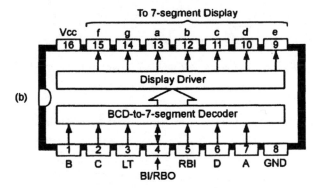

Figure 5.21 *Functional diagram of the (a) 74LS47 and (b) 74LS48 BCD-to-7-segment decoder/driver ICs*

latches. *Figure 5.21* shows the functional diagrams and pin designations of these devices, each of which is housed in a 16-pin DIL package.

The 74LS47 has active-low outputs designed for driving a common-anode LED display via external current-limiting resistors (Rx), as shown in *Figure 5.22*. The 74LS48 has active-high outputs designed for driving a common-cathode LED display in a manner similar to that of *Figure 5.22*, but with the display's common terminal taken to ground. The Rx resistors must limit the segment currents to less than 24mA in the 74LS47, and less than 6mA in the 74LS48. The 74LS48 can be used to drive a 7-segment LCD display by using the connections already shown in *Figure 5.11*.

Note from *Figure 5.21* that each of these ICs has three input 'control' terminals, these being designated LT (Lamp Test), BI/RBO, and RBI. The LT terminal drives all display outputs on when the terminal is driven to logic 0 with the RBO terminal open or at logic 1. When the BI/RBO terminal is pulled low all outputs are blanked; this pin also functions as a ripple-blanking output terminal. *Figure 5.23* shows how to connect the ripple-blanking

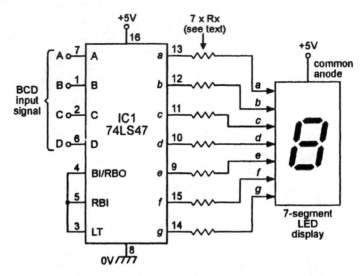

Figure 5.22   *Basic way of using a 74LS47 IC to drive a common-anode LED display*

terminals to give leading zero suppression on the first three digits of a 4-digit display.

The 4511B

The most popular CMOS 4000B-series BCD-to-7-segment LED-driving IC is the 4511B (also available as the 74HC4511), which has an integral 4-bit data latch but has no built-in ripple-blanking facilities. *Figure 5.24* shows the functional diagram and pin notations of the device, which can use any power

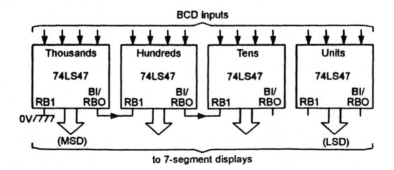

Figure 5.23   *Method of applying leading-zero suppression to the first three digits of a 4–digit display using 74LS47 ICs*

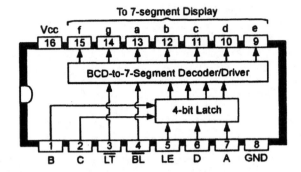

Figure 5.24 *Functional diagram and pin notations of the 4511B 7-segment latch/decoder/LED-driver IC*

source in the 5V to 15V range. The IC is ideally suited to driving common-cathode LED displays, and uses *npn* bipolar output transistor stages that can each source up to 25mA.

The 4511B is very easy to use, and has only three input control terminals; of these, the not-LT (pin-3) pin is normally tied high, but turns on all seven segments of the display when pulled low. The not-BL (pin-4) terminal is also normally tied high, but blanks (turns off) all seven segments when pulled low. Finally, the LE (latch enable) terminal (pin-5) enables the IC to give either direct or latched decoding operation; when LE is low, the IC gives direct decoding operation, but when LE is taken high it freezes the display.

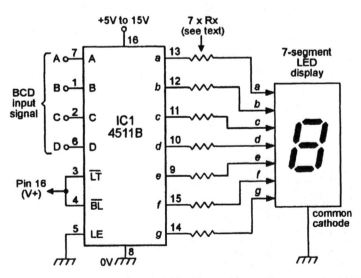

Figure 5.25 *Basic way of using the 4511B to drive a 7-segment common-cathode LED display*

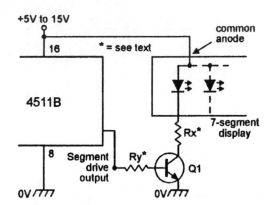

Figure 5.26   *Driving a common-anode LED display*

The 4511B can be used to drive most popular types of 7-segment display. *Figure 5.25* shows the basic connections for driving a common-cathode LED display; a current-limiting resistor (Rx) must be wired in series with each display segment and must have its value chosen to limit the segment current below 25mA.

Figures 5.26, 5.27 and *5.28* show how to modify the above circuit to drive LED common-anode displays, gas discharge displays, and low-brightness fluorescent displays respectively. Note in the cases of *Figures 5.26* and *5.27* that an *npn* buffer transistor must be interposed between each output drive segment and the input segment of the display; in each case, Rx determines the operating segment current of the display, and Ry determines the base current of the transistor.

The 4511B can also be used to drive 7-segment liquid crystal displays by using an external squarewave 'phase' signal and a set of EX-OR gates in a configuration similar to that of *Figure 5.11*. In practice, however, it is far better to use a 4543B IC for this particular application.

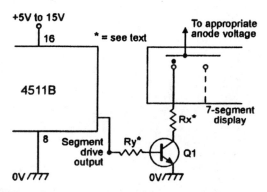

Figure 5.27   *Driving a gas discharge readout*

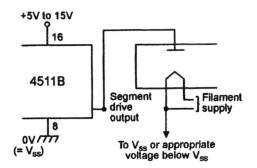

Figure 5.28 *Driving a low-bright-ness fluorescent readout*

The 4543B

The most popular 4000B-series BCD-to-7-segment LCD-driving IC is the 4543B (also available as the 74HC4543), which has a built-in data latch. *Figure 5.29* shows the IC's functional diagram and pin notations. The device incorporates an EX-OR array (of the type shown on *Figure 5.11*) in its output driver network, which can source or sink several milliamps of output current. This feature enables the IC to act as a universal unit that can drive common-cathode or common-anode LED or liquid crystal 7-segment displays with equal ease, as shown in *Figures 5.30* to *5.33*.

The 4543B has three input control terminals, these being designated not-LATCH, PHASE, and BL (BLANK). In normal use the not-LATCH terminal is biased high and the BL terminal is tied low. The state of the PHASE terminal depends on the type of display that is being driven. For driving LCD readouts, a squarewave (roughly 50Hz, swinging fully between the GND and Vcc values) must be applied to the PHASE terminal: for driving common-cathode LED displays, PHASE must be grounded: for driving common-anode displays, PHASE must be tied to logic high.

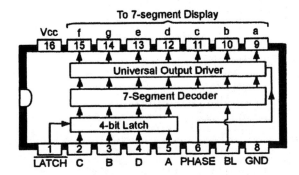

Figure 5.29 *Functional diagram and pin notations of the 4543B universal 7-segment latch/decoder/driver IC*

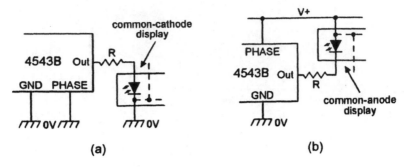

Figure 5.30   *Way of using the 4543B to drive (a) common-cathode or (b) common-anode 7-segment LED displays*

The display can be blanked at any time by driving the BL terminal to the logic-1 level. When the not-LATCH terminal is in its normal high (logic-1) state, BCD inputs are decoded and fed directly to the 7-segment output terminals of the IC. When the not-LATCH terminal is pulled low, the BCD input signals that are present at the moment of transition are latched into memory and fed (in decoded form) to the 7-segment outputs until the not-LATCH pin returns to the high state.

Figure 5.30 shows basic ways of using the 4543B to drive common-cathode and common-anode 7-segment LED displays; the 'R' resistance value must limit the output drive current to below 10mA per segment. *Figure 5.31* shows the basic way of using the 4543B to drive a 7-segment LCD, and *Figures 5.32* and *5.33* show it used to drive other types of 7-segment display; in *Figure*

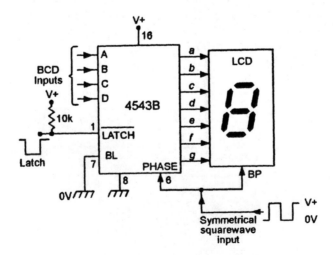

Figure 5.31   *Way of using the 4543B to drive a 7-segment LCD*

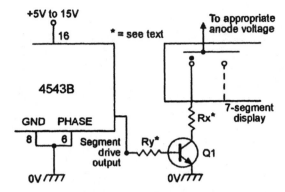

Figure 5.32 *Driving a gas discharge readout with a 4543B*

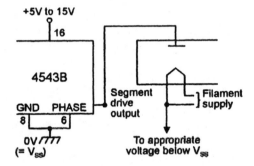

Figure 5.33 *Driving a fluorescent readout with a 4543B*

5.32, Rx sets the segment current of the display and Ry sets the base current of the transistor (10mA maximum).

The 4026B

The 4026B IC is a complete decade counter with integral decoder/driver circuitry that can directly drive a 7-segment common-cathode LED display. The segment output currents are internally limited to about 5mA at 10V or 10mA at 15V, enabling the display to be connected directly to the outputs of the IC without the use of external current-limiting resistors. The IC does not incorporate a data latch and has no facility for ripple blanking. *Figure 5.34* shows the functional diagram and pin notations of the 4026B.

The 4026B has four input control terminals, and three auxiliary output terminals. The input terminals are designated CLK (CLOCK), CLK INH (CLOCK INHIBIT), RESET, and DISPLAY ENABLE IN. The IC incor-

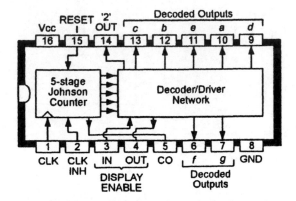

Figure 5.34 *Functional diagram and pin notations of the 4026B decade counter with 7-segment display driver*

porates a Schmitt trigger on its CLK input line, and clock signals do not have to be pre-shaped. The counter is reset to zero by driving the RESET terminal high.

The CLK INH terminal must be grounded to allow normal counting operation: when CLK INH is high the counters are inhibited. The display is blanked when the DISPLAY ENABLE IN terminal is grounded: the DISPLAY ENABLE IN terminal must be high for normal operation. Thus, in normal operation the RESET and CLK INH terminals are grounded and the DISPLAY ENABLE IN terminal is held positive, as shown in *Figure 5.35.*

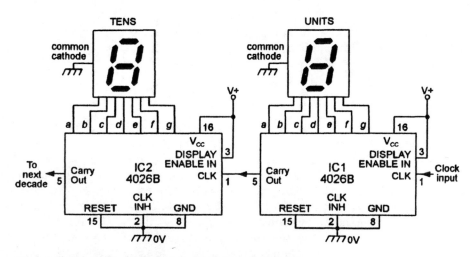

Figure 5.35 *Basic method of cascading 4026B ICs*

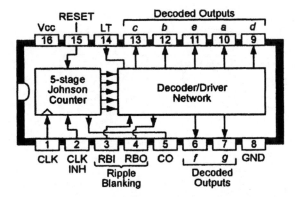

Figure 5.36 *Functional diagram and pin notations of the 4033B decade counter with 7-segment display driver*

The three auxiliary output terminals of the 4026B are designated DISPLAY ENABLE OUT, CO (CARRY OUT), and '2' OUT. The DISPLAY ENABLE OUT signal is a slightly delayed copy of the DISPLAY ENABLE IN input signal. The CO signal is a symmetrical squarewave at one-tenth of the CLK input frequency, and is useful in cascading 4026B counters. The '2' OUT terminal goes low only on a count of 2. *Figure 5.35* shows the basic circuit connections to be used when cascading stages.

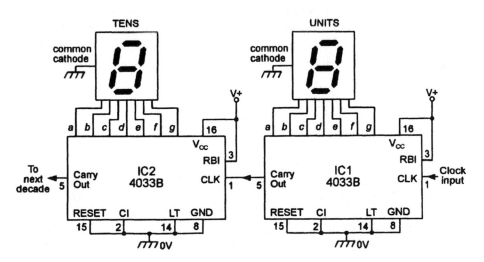

Figure 5.37 *Basic method of cascading 4033B ICs (without zero suppression)*

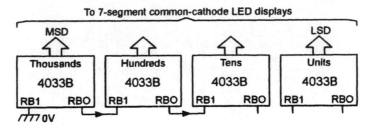

Figure 5.38   *Method of modifying the* Figure 5.37 *circuit to give automatic leading-zero suppression*

The 4033B

This device (see *Figure 5.36*) can be regarded as a modified version of the 4026B, with the DISPLAY ENABLE IN and DISPLAY ENABLE OUT terminals eliminated and replaced by ripple blanking input (RBI) and output (RBO) terminals, and with the '2' OUT terminal replaced with a LT (LAMP TEST) terminal which activates all output segments when biased high. In normal use the RESET, CLK INH and LT terminals are all grounded and the RBI terminal is made positive, as shown in *Figure 5.37*: this configuration does not provide blanking of unwanted leading and/or trailing zeros.

If cascaded 4033B ICs are required to give automatic leading-zero suppression the basic *Figure 5.37* circuit must be modified as shown in *Figure 5.38*, to provide ripple-blanking operation. Here, the RBI terminal of the most significant digit (MSD) is grounded, and its RBO terminal is connected to the RBI terminal of the next least-significant stage. This procedure is repeated on all except the LSD, which does not require zero suppression. If trailing-zero suppression is required, the direction of ripple-blanking feedback must be reversed, with the RBI terminal of the LSD grounded and its RBO terminal wired to the RBI terminal of the next least-significant stage, and so on.

6

Light-sensitive circuits

The previous four chapters have dealt with the theory and applications of light-generating and light-reflecting devices such as LEDs and LCDs. The present chapter concentrates on the operating principles and applications of light-sensitive devices such as LDRs and photodiodes, etc.

LDR basics

Electronic optosensors are devices that alter their electrical characteristics in the presence of visible or invisible light. The best known devices of these types are the LDR (light dependent resistor), the photodiode, the photo-transistor, and the PIR (passive infra-red) detector.

   LDR operation relies on the fact that the conductive resistance of a film of cadmium sulphide (CdS) varies with the intensity of light falling on the face of the film. This resistance is very high under dark conditions and low under bright conditions. *Figure 6.1* shows the LDR's circuit symbol and basic construction, which consists of a pair of metal film contacts separated by a snake-like track of light-sensitive cadmium sulphide film, which is designed to provide the maximum possible contact area with the two metal films. The structure is housed in a clear plastic or resin case, to provide free access to external light.

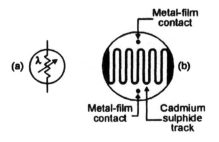

Figure 6.1 *LDR symbol (a) and basic structure (b)*

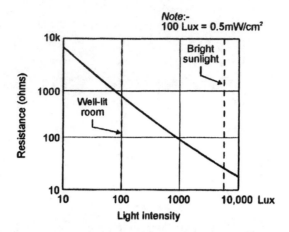

Figure 6.2 *Typical characteristics curve of a LDR with a 10mm face diameter*

Practical LDRs are available in a variety of sizes and package styles, the most popular size having a face diameter of roughly 10mm. *Figure 6.2* shows the typical characteristic curve of such a device, which has a resistance of about 900R at a light intensity of 100 Lux (typical of a well lit room) or about 30R at an intensity of 8000 Lux (typical of bright sunlight). The resistance rises to several megohms under dark conditions.

LDRs are sensitive, inexpensive, and readily available devices with power and voltage handling capabilities similar to those of conventional resistors. Their only significant defect is that they are fairly slow acting, taking tens or hundreds of milliseconds to respond to sudden changes in light level. Useful LDR applications include light- and dark-activated switches and alarms, light-beam alarms, and reflective smoke alarms, etc. *Figures 6.3* to *6.22* show some practical applications of the device; each of these circuits will work with virtually any LDR with a face diameter in the range 3mm to 12mm.

LDR light-switches

Figures 6.3 to *6.8* show some practical relay-output light-activated switch circuits based on the LDR. *Figure 6.3* shows a simple non-latching circuit, designed to activate when light enters a normally dark area such as the inside of a safe or cabinet, etc.

Here, R1–LDR and R2 form a potential divider that controls the base-bias of Q1. Under dark conditions the LDR resistance is very high, so negligible base-bias is applied to Q1, and Q1 and RLA are off. When a significant amount of light falls on the LDR face the LDR resistance falls to a fairly low value and base-bias is applied to Q1, which thus turns on and activates

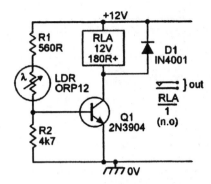

Figure 6.3 *Simple non-latching light-activated relay switch*

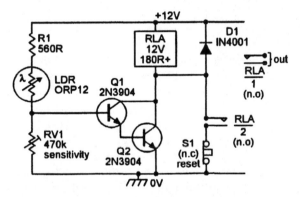

Figure 6.4 *Sensitive self-latching light-activated relay switch*

the RLA/1 relay contacts, which can be used to control external circuitry. The relay can be any 12V type with a coil resistance of 180R or greater.

The simple *Figure 6.3* circuit has a fairly low sensitivity, and has no facility for sensitivity adjustment. *Figure 6.4* shows how these deficiencies can be overcome by using a super-alpha-connected pair of transistors in place of Q1, and by using sensitivity control RV1 in place of R2; this circuit can be activated by LDR resistances as high as 200k (i.e. by exposing the LDR to very small light levels), and draws a standby current of only a few microamps under 'dark' conditions. The diagram also shows how the circuit can be made to give a self-latching action via relay contacts RLA/2; normally-closed push-button switch S1 enables the circuit to be reset (unlatched) when required.

Figure 6.5 shows how a LDR can be used to make a simple dark-activated relay switch that turns RLA on when the light level falls below a value preset via RV1. Here, R1 and LDR form a potential divider that generates an output voltage that rises as the light level falls. This voltage is buffered by

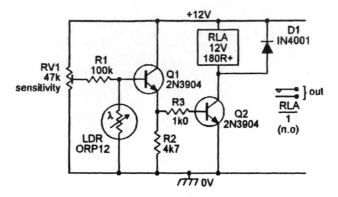

Figure 6.5 *Simple dark-activated relay switch*

emitter-follower Q1 and used to control relay RLA via common-emitter amplifier Q2 and current-limiting resistor R3.

The light trigger points of the *Figure 6.3* to *6.5* circuits are susceptible to variations in circuit supply voltage and ambient temperature. *Figure 6.6* shows a very sensitive precision light-activated circuit that is not influenced by such variations. In this case, LDR–RV1 and R1–R2 are connected in the form of a Wheatstone bridge, and the op-amp and Q1–RLA act as a highly sensitive balance-detecting switch. The bridge balance point is quite independent of variations in supply voltage and temperature, and is influenced only by variations in the relative values of the bridge components.

In *Figure 6.6*, the LDR and RV1 form one arm of the bridge, and R1–R2 form the other arm. These arms act as potential dividers, with the R1–R2 arm applying a fixed half-supply voltage to the non-inverting input of the op-amp, and with the LDR–RV1 divider applying a light-dependent variable voltage to the inverting terminal of the op-amp.

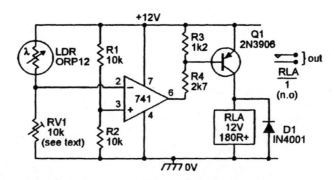

Figure 6.6 *Precision light-sensitive relay switch*

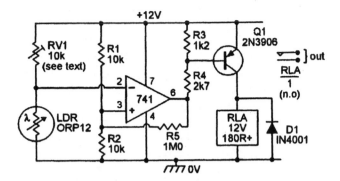

Figure 6.7 *Precision dark-activated switch, with hysteresis*

In use, RV1 is adjusted so that the LDR–RV1 voltage rises slightly above that of R1–R2 as the light intensity rises to the desired trigger level, and under this condition the op-amp output switches to negative saturation and thus drives the relay on via Q1 and biasing resistors R3–R4. When the light intensity falls below this level, the op-amp output switches to positive saturation, and under this condition Q1 and the relay are off.

The *Figure 6.6* circuit is very sensitive and can detect light-level changes too small to be seen by the human eye. The circuit can be modified to act as a precision dark-activated switch by either transposing the inverting and non-inverting input terminals of the op-amp, or by transposing RV1 and the LDR. *Figure 6.7* shows a circuit using the latter option.

Figure 6.7 also shows how a small amount of hysteresis can be added to the circuit via feedback resistor R5, so that the relay turns on when the light level falls to a particular value, but does not turn off again until the light intensity rises a substantial amount above this value. The magnitude of

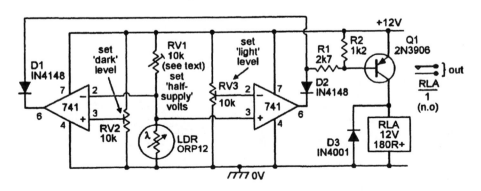

Figure 6.8 *Combined light-/dark-activated switch with single relay output*

hysteresis is inversely proportional to the R5 value, being zero when R5 is open-circuit.

A precision combined light/dark switch, which activates a single relay if the light intensity rises above one pre-set value or falls below another pre-set value, can easily be made by combining an op-amp light switch and an op-amp dark switch in the manner shown in *Figure 6.8*.

To set up the *Figure 6.8* circuit, first trim RV1 so that approximately half-supply volts appear on the LDR–RV1 junction when the LDR is illuminated at the mean or normal intensity level. RV2 can then be pre-set so that RLA turns on when the light intensity falls to the desired dark level, and RV3 can be adjusted so that RLA activates at the desired brightness level.

Note in *Figures 6.6* to *6.8* that the adjusted RV1 value should equal the LDR resistance value at the normal light level of each circuit.

Bell-output LDR alarms

The *Figure 6.3* to *6.8* light-activated LDR circuits all have relay outputs that can be used to control virtually any type of external circuitry. In some light-activated applications, however, circuits are required to act as audible-output alarms, with a bell or siren-sound output, and this type of action can be obtained without the use of relays. *Figures 6.9* to *6.11* show three practical 'alarm bell output' circuits of this type.

The *Figure 6.9* to *6.11* circuits are each designed to give a direct output to an alarm bell, which must be of the self-interrupting type and must consume an operating current of less than 2A. The supply voltage of each circuit should be 1.5V to 2V greater than the nominal operating value of the bell.

Figure 6.9 shows the circuit of a simple light-activated alarm; the operating theory is fairly simple. The R1–LDR and R2 form a potential divider: under dark conditions the LDR resistance is high, so the LDR–R2 junction voltage is too small to activate the gate of the SCR (silicon controlled recti-

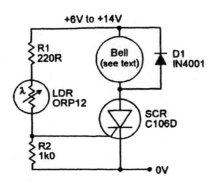

Figure 6.9   *Simple light-activated alarm bell*

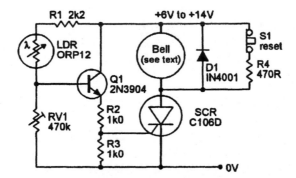

Figure 6.10 *Improved light-activated alarm bell with self-latching facility*

fier), but under bright conditions the LDR resistance is low, so gate bias is applied to the SCR, which turns on and activates the alarm bell.

Note in the above circuit that, although the SCR is a self-latching device, the fact that the bell is of the self-interrupting type ensures that the SCR automatically unlatches repeatedly as the bell operates (and the SCR anode current falls to zero in each self-interrupt phase). Consequently, the alarm bell automatically turns off again when the light level falls back below the trip level.

The *Figure 6.9* circuit has a fairly low sensitivity and has no facility for sensitivity adjustment. *Figure 6.10* shows how these weaknesses can be overcome by using RV1 in place of R2 and by using Q1 as a buffer between the LDR and the SCR gate. This diagram also shows how the circuit can be made self-latching by wiring R4 across the bell so that the SCR anode current does not fall to zero as the bell self-interrupts. Switch S1 enables the circuit to be reset (unlatched) when required.

Figure 6.11 shows how to make a precision light-activated alarm with a SCR-driven alarm bell output by using a Wheatstone bridge

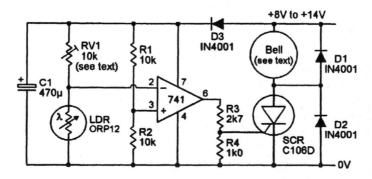

Figure 6.11 *Precision light-activated alarm bell*

(LDR–RV1–R1–R2) and an op-amp balance detector to give the precision action (as described in the basic *Figure 6.6* circuit). This circuit can be converted into a dark-activated alarm by simply transposing RV1 and the LDR. Hysteresis can also be added, if required.

Siren-output LDR alarms

Figures 6.12 to *6.17* show ways of using CMOS 4001B quad 2-input NOR gate ICs as the basis of various light-activated 'siren-sound' alarms that generate audible outputs in loudspeakers. The *Figure 6.12* circuit is that of a light-activated alarm that generates a low-power (up to 520mW) 800Hz pulsed-tone signal in the speaker when the light input exceeds a pre-set threshold value. Here, IC1c and IC1d are wired as an 800Hz astable multivibrator that can feed tone signals into the speaker via Q1 and is gated on only when the output of IC1b is low, and IC1a–IC1b are wired as a 6Hz astable that is gated on only when its pin-1 gate terminal (which is coupled to the LDR–RV1 potential divider) is pulled low.

The action of the *Figure 6.12* circuit is as follows. Under dark conditions the LDR–RV1 junction voltage is high, so both astables are disabled and no signal is generated in the speaker. Under 'light' conditions the LDR–RV1 junction voltage is low, so the 6Hz astable is activated and in turn gates the 800Hz astable on and off at a 6Hz rate, thereby generating a pulsed-tone signal in the speaker via Q1.

The precise switching or gate point of the 4001B IC is determined by the threshold voltage value of the IC, and this is a percentage value of the supply voltage: the value is nominally 50%, but may vary from 30% to 70% between individual ICs. In practice, the switching point of each individual 4001B IC

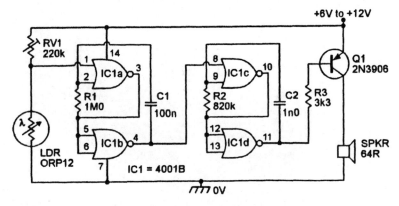

Figure 6.12 *Light-activated alarm with pulsed-tone output*

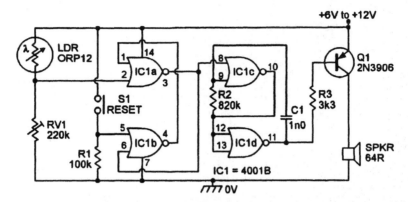

Figure 6.13 *Self-latching light-activated alarm with monotone output*

is very stable, and the *Figure 6.12* circuit gives very sensitive 'light'-activated alarm triggering.

Figure 6.13 shows the circuit of a self-latching light-activated alarm with a 800Hz monotone output. In this case IC1c–IC1d are again wired as a gated 800Hz astable, but IC1a–IC1b are wired as a bistable multivibrator with an output that (under dark conditions) is normally high, thus gating the 800Hz astable off. Under bright conditions, however, the LDR–RV1 junction goes high and latches the bistable into its alternative 'output low' state, thereby gating the 800Hz astable on and generating the monotone alarm signal; once latched, the circuit remains in this 'on' state until dark conditions return and the bistable is simultaneously reset via S1.

Note that the light/dark operation of the *Figure 6.12* and *6.13* circuits can be reversed by simply transposing the LDR–RV1 positions. The sensitivity levels of these two basic circuits are adequate for most practical purposes, but can, if required, be boosted (and the trigger-level stability increased), by interposing an op-amp voltage comparator (of the basic *Figure 6.6* or *6.7* type) between the LDR–RV1 light-sensitive potential divider and the gate terminal of the CMOS waveform generator, as shown in the *Figure 6.14* circuit; resistor R3 controls the hysteresis of the circuit, and can be removed if the hysteresis is not needed.

The basic *Figure 6.12* to *6.14* circuits generate fairly modest values of acoustic output power, with the power input to the 64 ohm loudspeaker reaching a maximum value of 520mW when using a 12V supply. If desired, the input power to the loudspeaker can be boosted to well over a dozen watts by using more complex output-driving circuitry, as shown in *Figures 6.15* to *6.17*. Note in all these circuits that maximum *acoustic* output powers can be obtained by using cheap horn-type loudspeakers, which typically have electro-acoustic power conversion efficiencies that are twenty to forty times greater than normal hi-fi speakers.

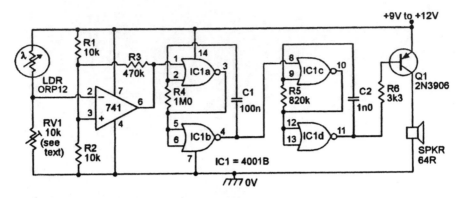

Figure 6.14 *Precision light-activated pulsed-tone alarm with hysteresis*

Figure 6.15 shows the type of output stage that is used in the basic *Figure 6.12* to *6.14* circuits. Thus, when the siren waveform generator is gated off its output is high and Q1 is thus cut off, but when the generator is gated on its output drives Q1 on and off and causes it to feed power to the 64R speaker. The output power depends on the supply rail voltage, and has a value of about 520mW at 12V, or 120mW at 6V, when feeding a 64R speaker

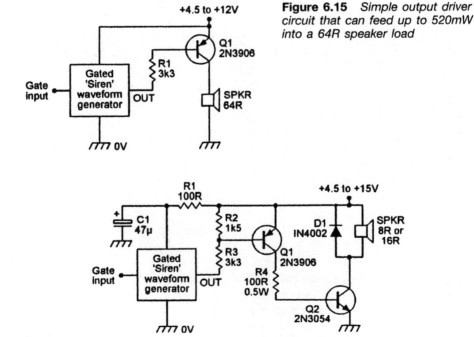

Figure 6.15 *Simple output driver circuit that can feed up to 520mW into a 64R speaker load*

Figure 6.16 *Medium power (up to 6.6 watts into 8R0) output driver*

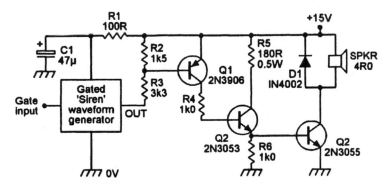

Figure 6.17 *High power (up to 13.2 watts into 4R0) output driver*

load. Note that, since Q1 is used as a simple power switch in this application, very little power is lost across the 2N3906 transistor, but its current rating (200mA maximum) may be exceeded if the circuit is used with a supply value greater than 12V.

Figure 6.16 shows how the basic *Figure 6.15* circuit can be modified so that it can feed a maximum of 6.6 watts of audio power into an 8R0 speaker load, or 3.3 watts into a 16R load. Here, both transistors are cut off when the waveform generator is gated off, but are switched on and off in sympathy with the siren waveform when the generator is gated on. Note in this circuit that the positive power supply rail is fed directly to the output driver, but is fed to the waveform generator via decoupling network R1–C1, that voltage divider R2–R3 ensures that the output stages are not driven on until the generator's output voltage falls at least 1.9V below the supply rail value, and that diode D1 is used to damp the speaker's back-e.m.f. when driver Q2 switches off.

Finally, the *Figure 6.17* driver circuit can pump a maximum of 13.2 watts into a 4R0 speaker load when powered from a 15V supply. Here, all three transistors are cut off when the waveform generator is cut off, but are switched on and off in sympathy with the siren waveform when the generator is gated on.

Miscellaneous LDR circuits

One of the best-known types of LDR application is in light-beam alarms or switches. A light-beam alarm system consists of a focused light-beam transmitter (Tx) and a focused light-beam receiver (Rx), and may be configured to give either a direct-light-beam or a reflected-light-beam type of optical contact operation. *Figure 6.18* shows the basic elements of a direct-light-beam type of alarm system, in which the sharply focused Tx light-beam is

Figure 6.18   *The basic elements of a direct-light-beam type of alarm system*

aimed directly at the light-sensitive input point of the Rx unit, which (usually) is designed to activate an external alarm or safety mechanism if a person, object, or piece of machinery enters the light-beam and breaks the optical contact between the Tx and Rx.

Figure 6.19 shows a very simple example of a lamp-and-LDR direct-light-beam system that activates an alarm-bell if the beam is interrupted. The beam is generated via an ordinary electric lamp and a lens, and is focused (via a 'collector' lens) on to the face of an LDR in the remote Rx unit, which operates as a dark-activated alarm. Normally, the LDR face is illuminated by the light-beam, so the LDR has a low resistance and very little voltage thus appears on the RV1–LDR junction, so the SCR and bell are off. When the light-beam is broken, however, the LDR resistance goes high and enough voltage appears on the RV1–LDR junction to trigger the SCR, which drives the alarm bell on; R3 is used to self-latch the alarm.

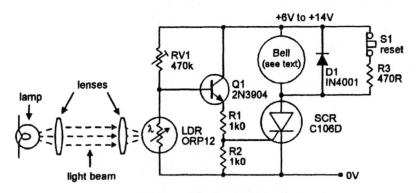

Figure 6.19 *Simple direct-light-beam alarm with alarm bell output*

Figure 6.20 shows the basic elements of a reflected-light-beam type of alarm system, in which the Tx light-beam and Rx lens are optically screened from each other but are both aimed outwards towards a specific point, so that an optical link can be set up by a reflective object (such as metallic paint

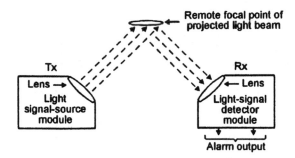

Figure 6.20 *The basic elements of a reflected-light-beam type of alarm system*

or smoke or fog particles) placed at that point. This type of system is usually designed to activate an alarm when the presence of such an object is detected, but can also be configured to give the reverse action, so that the alarm activates if a reflective object is illegally removed.

Units of the *Figure 6.20* type were once widely used in conjunction with reflection-type fog and smoke detector units; *Figure 6.21* shows a sectional view of a smoke detector unit of this type. Here, the lamp and LDR are mounted in an open ended but light-excluding box, in which an internal screen prevents the lamp-light from falling directly on the LDR face. The lamp is a source of both light and heat, and the heat causes convection currents of air to be drawn in from the bottom of the box and to be expelled through the top. The inside of the box is painted matt black; its construction lets air pass through the box but excludes external light.

Thus, if the convected air currents are smoke free, no light falls on the LDR face, and the LDR presents a high resistance. If the air currents do contain smoke, however, the smoke particles cause the light of the lamp to reflect on to the LDR face and so cause a large and easily detectable decrease in the LDR resistance. *Figure 6.22* shows a reflection-type smoke alarm circuit that can be used with this detector; the circuit acts in the same way as the improved *Figure 6.10* light-activated alarm circuit.

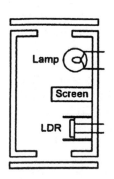

Figure 6.21 *Sectional view of reflection-type smoke detector*

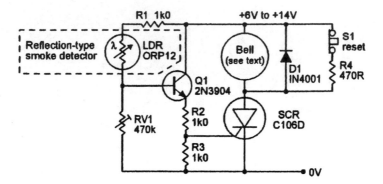

Figure 6.22   *Smoke alarm with alarm bell output*

Simple lamp-and-LDR light-beam alarms of the types described in this section have several obvious disadvantages in most modern security-alarm applications. Their light-beams are, for example, clearly visible to an intruder, the transmitter's filament lamp is unreliable, and the systems are (because the transmitter's filament lamp wastes a lot of power) very ineffi- cient. In practice, virtually all modern light-beam security systems operate in the invisible infra-red (rather than visible light) range, use one or more pulse- driven IR LEDs to generate the transmitter's 'light-beam', and use match- ing IR photodiodes or phototransistors to detect the beam at the receiver end of the system. A number of practical circuits of these types are presented in Chapter 11 of this book.

Photodiodes

Cadmium sulphide (CdS) LDRs are sensitive but slow-acting light sensors. Generally, they are ideal for use in slow-acting direct-coupled light-level sensing applications, but are quite unsuitable for use as optical sensors in medium- to high-speed applications. The ideal optical sensors for use in the latter types of application are the silicon photodiode and the silicon phototransistor.

In its very crudest form, a photodiode is a normal silicon diode minus its opaque (light-excluding) covering. If a normal silicon diode is connected in the reverse-biased circuit of *Figure 6.23*, negligible current flows through the diode and zero voltage is developed across R1. If the diode's opaque cover- ing is now removed (so that the diode's semiconductor junction is revealed) and the junction is then exposed to visible light in the same circuit, the diode will pass a significant reverse current and thus generate an output voltage across R1. The magnitude of the reverse current and the output voltage is directly proportional to the intensity of the light source, and the diode is thus truly photosensitive.

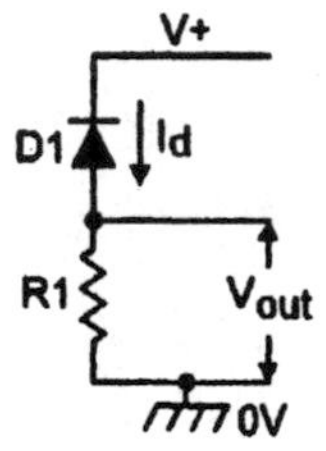

Figure 6.23 *Reverse-biased diode circuit*

All silicon junctions are photosensitive, and a basic photodiode can – for most practical purposes – be regarded as a normal diode housed in a case that lets external light easily reach its photosensitive semiconductor junction. *Figure 6.24(a)* shows the standard photodiode symbol. In use, the photodiode is reverse biased and the output voltage is taken from across a series-connected load resistor; this resistor can be connected between the diode and ground, as in *Figure 6.24(b)*, or between the diode and the positive supply line, as in *Figure 6.24(c)*.

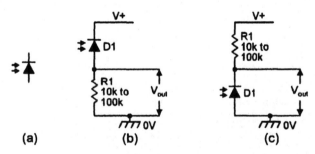

(a) (b) (c)

Figure 6.24 *Photodiode symbol (a) and alternative ways ((b) and (c)) of using a photodiode as a light-to-voltage converter*

In reality, the physical form of a normal silicon diode's *pn* junction is such that the device exhibits fairly low optical sensitivity; all practical photodiodes use special types of junction design to maximize their effective photosensitivity. Most photodiodes come in one or other of two basic types, being either 'simple' photodiodes or PIN photodiodes. *Figure 6.25* illustrates some basic points on these subjects.

Normal silicon junction diodes use the basic form of construction shown (in symbolic form) in *Figure 6.25(a)*, in which the device's *p*- and *n*-type materials are moderately thick (and thus fairly opaque) and are effectively fused directly together to form the device's junction; the relatively high opacity of the *pn* junction's materials gives the junction fairly poor photosensitivity. In a simple photodiode, the photosensitivity is greatly increased by using a very thin (and thus highly translucent) slice of material on the *p*-type side of the junction, as shown in *Figure 6.25(b)*; external light can be

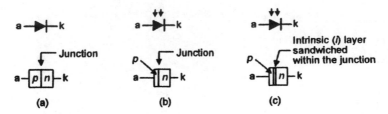

Figure 6.25   *Symbol and basic construction of (a) a normal silicon junction diode, (b) a simple photodiode, and (c) a PIN photodiode*

applied, via a built-in lens or window, to the optosensitive *pn* junction via this thin slice of *p*-type material.

Simple *Figure 6.25(b)*-type photodiodes have minimum on/off switching times of about 1μs, and can thus be used at maximum pulsed or switched operating frequencies of about 300kHz. The prime cause of this relatively long switching time is the high capacitance that occurs at the device's junction, between the *p*- and *n*-type materials. This problem is greatly reduced in PIN photodiodes, in which a very thin slice of intrinsic ('I') or 'undoped' silicon material is interposed at the junction between the *p*- and *n*-type materials, as shown in *Figure 6.25(c)*, thus greatly reducing the *p*-to-*n* junction's capacitance value.

Modern PIN-type photodiodes have typical minimum on/off switching times of about 10ns, and can thus be used at maximum switched-mode operating frequencies of about 30MHz, which is adequate for the vast majority of practical optoelectronic applications (in cases where even higher

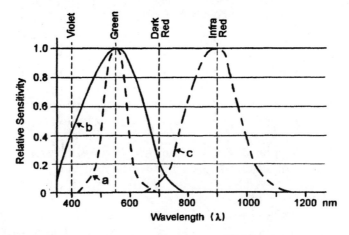

Figure 6.26   *Typical spectral response curves of (a) the human eye and (b) general-purpose and (c) IR photodiodes*

switching frequency optical sensing is required, special ultra-high-frequency avalanche-type photodiodes can be used).

Photodiodes can be designed to respond to either visible light or to IR light. The human eye has the type of spectral response curve shown in curve '*a*' in *Figure 6.26*. It has a maximum sensitivity to the colour green, which has a wavelength of about 550nm, but has a relatively low sensitivity to the colour violet (400nm) at one end of the spectrum and to dark red (700nm) at the other. General-purpose visible-light photodiodes have typical spectral response characteristics like those shown in curve '*b*' in *Figure 6.26*, and infra-red (IR) types have the type of response shown in curve '*c*'.

Phototransistors

Ordinary silicon transistors are made from an *npn* or *pnp* sandwich, and thus inherently contain a pair of photosensitive junctions. Some types are available in phototransistor form, and use the standard symbol shown in *Figure 6.27(a)*. *Figures 6.27(b)* to *6.27(d)* show three basic ways of using a phototransistor; in each case the base–collector junction is effectively reverse biased and thus acts as a photodiode. In *(b)* the base is grounded, and the transistor acts as a simple photodiode. In *(c)* and *(d)* the base terminal is open-circuit and the photogenerated currents effectively feed directly into the base and, by normal transistor action, generate a greatly amplified collector-to-emitter current that produces an output voltage across series resistor R1.

The sensitivity of a phototransistor is typically one hundred times greater than that of a photodiode, but its useful maximum operating frequency (usually a few hundred kilohertz) is proportionally lower than that of a photodiode. Most phototransistors are manufactured in 2-pin form, with only the device's collector and emitter made externally available; 3-pin types can be used in any of the basic configurations shown in *Figure 6.27*. Some phototransistors are made in very-high-gain Darlington form.

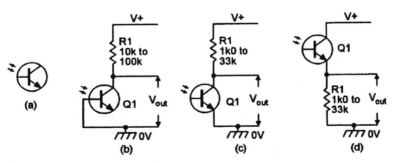

Figure 6.27 *Phototransistor symbol (a) and alternative ways (b) to (d) of using a phototransistor*

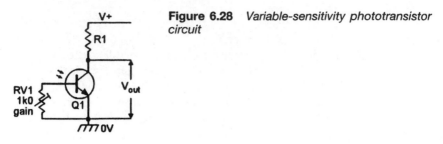

Figure 6.28 *Variable-sensitivity phototransistor circuit*

The sensitivity (and operating speed) of a 3-pin phototransistor can be made variable by wiring a variable resistor between its base and emitter, as shown in *Figure 6.28*; with RV1 open circuit, phototransistor operation is obtained; with RV1 short circuit, photodiode operation occurs.

Note in the *Figure 6.24, 6.27,* and *6.28* photodiode and phototransistor circuits that, in practice, the R1 load value is usually chosen on a compromise basis, since the circuit sensitivity increases but the useful operating bandwidth decreases as the R1 value is increased. Also, the R1 value must, in many applications, be chosen to bring the photosensitive device into its linear operating region.

IR pre-amp circuits

Photodiodes or phototransistors are often used as the sensing elements at the receiver end of light-beam alarms, remote-control, or fibre optic cable systems. In such applications, the signal reaching the photosensor may vary considerably in strength, and the sensor may be subjected to a great deal of noise in the form of unwanted visible or IR light signals, etc. To help minimize these problems, the systems are usually operated in the IR range, and the optosensor output is passed to processing circuitry via a low-noise

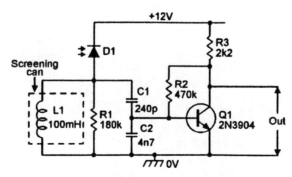

Figure 6.29 *Selective IR pre-amplifier designed for 30kHz operation*

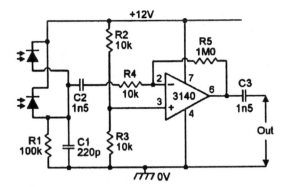

Figure 6.30 *20kHz selective pre-amp for use in IR light-beam applications*

pre-amplifier with a wide dynamic operating range. *Figures 6.29* and *6.30* show typical examples of such circuits, using photodiode sensors.

The *Figure 6.29* circuit is designed to detect an IR optical signal that is switched at a 30kHz rate. Photodiode D1 senses the IR signal and feeds it into 30kHz tuned circuit L1–C1–C2, which is lightly damped by R1. The resulting frequency-selected low-noise output of the tuned circuit is tapped off at the C1–C2 junction and then amplified by Q1.

Finally, *Figure 6.30* shows a 20kHz selective pre-amplifier circuit for use in an IR light-beam alarm application in which the alarm sounds when the beam is broken. Here, two IR photodiodes are wired in parallel (so that beam signals are lost only when *both* diode signals are cut off) and share a common 100k load resistor (R1). This resistor is shunted by C1 to reject unwanted high-frequency signals, and the R1 output signals are fed to the ×100 op-amp inverting amplifier via C2, which rejects unwanted low-frequency signals.

PIR movement-detecting systems

IR light-beam alarms are active IR units that react to an artificially generated source of IR radiation. Passive IR (PIR) alarms, on the other hand, react to naturally generated IR radiation such as the heat-generated IR energy radiated by the human body, and are widely used in modern security systems. Most PIR security systems are designed to activate an alarm or floodlight, or open a door or activate some other mechanism, when a human or other large warm-blooded animal moves about within the sensing range of a PIR detector unit, and use a pyroelectric IR detector of the type shown in *Figure 6.31* as their basic IR-sensing element.

The basic *Figure 6.31* pyroelectric IR detector makes use of special ceramic elements that generate electrical charges when subjected to thermal varia-

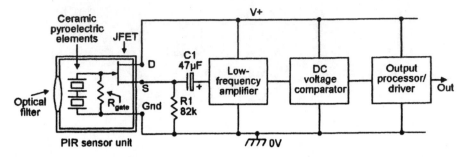

Figure 6.31 *Basic PIR detector usage circuit*

tions or uneven heating. Modern pyroelectric IR detectors such as the popular PIS201S and E600STO types incorporate two small opposite-polarity series-connected ceramic elements of this type, with their combined output buffered via a JFET source-follower, and have the IR input signals focused onto the ceramic elements by a simple filtering lens, as shown in the basic PIR detector usage circuit of *Figure 6.31*. It is important to note at this point that the detector's final output voltage is proportional to the *difference* between the output voltages of the two ceramic elements.

The basic action of the *Figure 6.31* PIR detector is such that, when a human body is within the visual field of the pyroelectric elements, part of that body's radiated IR energy falls on the surfaces of the elements and is converted into a small but detectable variation in surface temperature and corresponding variation in the output voltage of each element. If the human body (or other source of IR radiation) is stationary in front of the detector's lens under this condition, the two elements generate identical output voltages and the unit's final 'difference' output is thus zero, but if the body is moving while in front of the lens the two elements generate different output voltages and the unit produces a varying output voltage.

Thus, when the PIR unit is wired as shown in the *Figure 6.31* basic usage circuit, this movement-inspired voltage variation is made externally available via the buffering JFET and dc-blocking capacitor C1 and can, when suitably amplified and filtered, be used to activate an alarm or other mechanism when a human body movement is detected. In practice, pyroelectric IR detectors of the simple type just described have, because of the small size (usually about 20mm²) and simple design of the detector's IR-gathering lens, maximum useful detection ranges of roughly one metre. In modern commercial PIR movement detecting security units, however, this range is usually extended to at least ten metres with the aid of a large (about 2000mm²) multi-faceted external IR-gathering/focusing plastic lens, which splits the visual field into a number of parallel strips and focuses them onto the two sensing areas of the PIR unit.

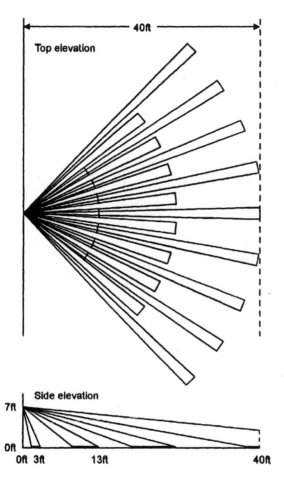

Figure 6.32 shows the typical PIR sensing pattern of a commercial 'intrusion detector' unit designed to protect a normal-sized room in domestic-type applications. In this example the unit is mounted on a wall at a height of seven feet and is aimed downwards at a shallow angle, and the multi-faceted plastic lens splits the visual field into a large number of vertical and horizontal segments. Any person moving through a single segment will activate a single trigger signal within the PIR sensor; a person moving through the entire visual field will thus produce numerous triggering signals, but a stationary IR source will produce no signals. Most intrusion detectors of this type incorporate 'event counting' circuitry that will only generate an alarm-activating output if three or more trigger signals are detected within a few seconds, thus minimizing the chances of a false alarm due to sudden changes in temperature caused by the auto-activation of time-switched security lights, etc.

The lens-generated PIR sensor pattern shown in *Figure 6.32* is the type usually used to protect single rooms in domestic burglar-alarm systems. Alternative lenses offer different ranges and coverage patterns for various special types of application; among the most important of these are the 'pet' type, in which the field's vertical span is restricted to 2.5 to 6.6 feet above ground level to avoid activation by domestic pets while giving good sensitivity to normal humans, and the 'corridor' type, in which the field's horizontal span is restricted to about 20 degrees to give long-distance coverage (typically about 30 metres) of narrow corridors and passageways.

Note that, because high-quality commercial PIR security units of this basic type are widely available at comparatively low cost, it is not practicable (on aesthetic and cost-effective grounds) to try to build similar units on a DIY basis.

7

Optocoupler circuits

Previous chapters have dealt with light-generating devices such as LEDs, and with light-sensitive devices such as phototransistors. The present chapter deals with so-called 'optocoupler' devices, which incorporate both a LED and a light-sensitive device in a single package. Optocoupler devices have a wide range of practical applications.

Optocoupler basics

An optocoupler device can be simply described as a sealed self-contained unit that houses independently powered optical (light) Tx and Rx units that can be coupled together optically. *Figure 7.1* shows the basic form of such a device. Here, the Tx unit is a LED, but the Rx unit may take the form of a phototransistor, a photo-FET, an opto-triac, or some other type of photo-sensitive semiconductor element; the Tx and Rx units are housed closely together in a single sealed package.

Most modern optocoupler devices use a phototransistor as their Rx unit; such a device is known simply as an 'optocoupler', since the input (the LED)

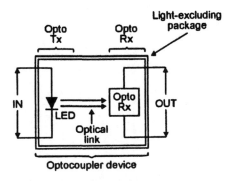

Figure 7.1 *Basic form of an optocoupler device*

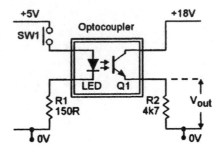

Figure 7.2   *Basic form and application circuit of a typical optocoupler*

and the output (the phototransistor) devices are optically coupled. *Figure 7.2* shows the basic form of an optocoupler, together with a very simple application circuit. Here, when SW1 is open no current flows in the LED, so no light falls on the face of Q1; Q1 passes virtually zero collector current under this condition, so zero voltage is developed across output resistor R2. Alternatively, when SW1 is closed, current flows through the LED via R1, and the resulting light falls on Q1 face, causing the phototransistor to conduct and generate an output voltage across R2.

Major points to note about the *Figure 7.2* optocoupler are that its output current is controlled by its input current, that a control circuit connected to its input can be electrically fully isolated from the output circuit, and that – since the input controls the output via a purely optical link – potential differences of hundreds of volts can safely exist between the input and output circuits without adversely influencing the optocoupler action. This 'isolating' characteristic is the main attraction of this type of optocoupler, which is generally known as an isolating optocoupler.

The simple application circuit of *Figure 7.2* can be used with digital input/output signals only, but in practice this basic circuit can easily be modified for use with analogue input/output signals, as shown later in this chapter. Typical isolating optocoupler applications include low-voltage to high-voltage (or vice versa) signal coupling, interfacing of a computer's output signals to external electronic circuitry or electric motors, etc., and interfacing of ground-referenced low-voltage circuitry to floating high-voltage circuitry driven directly from the mains AC power lines, etc. Optocouplers can also be used to replace low-power relays and pulse transformers in many applications.

Special optocouplers

The *Figure 7.2* device is a simple isolating optocoupler. *Figure 7.3* and *7.4* show two other types of optocoupler. The device shown in *Figure 7.3* is

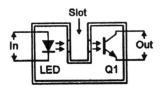

Figure 7.3 *Slotted optocoupler device*

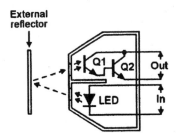

Figure 7.4 *Reflective optocoupler*

known as a slotted optocoupler, and has a slot moulded into the package between the LED light source and the phototransistor light sensor. Here, light can normally pass from the LED to Q1 without significant attenuation by the slot. The optocoupling can, however, be completely blocked by placing an opaque object in the slot. The slotted optocoupler can thus be used in a variety of 'presence' detecting applications, including end-of-tape detection, limit switching, and liquid-level detection.

The device shown in *Figure 7.4* is known as a reflective optocoupler. Here, the LED and Q1 are optically screened from each other within the package, and both face outwards (towards a common point) from the package. The construction is such that an optocoupled link can be set up by a reflective object (such as metallic paint or tape, or even smoke particles) sited a short distance outside the package, in line with both the LED and Q1. The reflective optocoupler can thus be used in applications such as tape-position detection, engine-shaft revolution counting or speed measurement, or smoke or fog detection, etc.

Optocoupler transfer ratios

One of the most important parameters of an optocoupler device is its optocoupling efficiency, and to maximize this parameter the LED and the phototransistor (which usually operate in the infra-red range) are always closely matched spectrally.

The most convenient way of specifying optocoupling efficiency is to quote the output-to-input current transfer ratio (CTR) of the device, i.e. the ratio of the output collector current (I_C) of the phototransistor, to the forward current (I_F) of the LED. Thus, CTR = I_C/I_F. In practice, CTR may be expressed as a simple figure such as 0.5, or (by multiplying this figure by 100) as a percentage figure such as 50%.

Simple isolating optocouplers with single-transistor output stages have typical CTR values on the range 20% to 100%; the actual CTR value

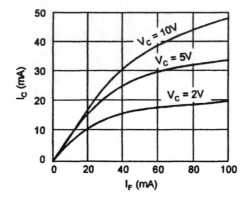

Figure 7.5 *Typical I_C/I_F characteristics of a simple optocoupler at various values of output transistor collector voltage (V_C)*

depends (among other things) on the input and output current values of the device and on the supply voltage value (V_C) of the phototransistor. *Figure 7.5* shows three typical sets of output/input currents obtained at different V_C values.

It should be noted that, because of variations in LED radiation efficiency and phototransistor current gains, the actual CTR values of individual optocouplers may vary significantly from the typical value. An optocoupler type with a typical CTR value of 60% may, for example, in fact have a true value in the range 30% to 90% in an individual device.

Other parameters

Other important optocoupler parameters include the following.

Isolation voltage. This is the maximum permissible DC potential that can be allowed to exist between the input and output circuits. Typical values vary from 500V to 4kV.

$V_{CE}(MAX)$. This is the maximum allowable DC voltage that can be applied across the output transistor. Typical values vary from 20V to 80V.

$I_F(MAX)$. This is the maximum permissible DC current that can be allowed to flow in the input LED. Typical values vary from 40mA to 100mA.

Bandwidth. This is the typical maximum signal frequency that can be usefully passed through the optocoupler when the device is operated in its normal mode. Typical values vary from 20kHz to 500kHz, depending on the type of device construction.

Practical optocouplers

Optocouplers are produced by several manufacturers and are available in a variety of forms and styles. Simple optocouplers are widely available in six basic forms, which are illustrated in *Figures 7.6* to *7.8*. Four of these (*Figures 7.6* and *7.7*) are isolating optocouplers, and the remaining two are the slotted optocoupler (*Figure 7.8(a)*) and the reflective optocoupler (*Figure 7.8(b)*). The table of *Figure 7.9* lists the typical parameter values of these six devices.

The simple isolating optocoupler (*Figure 7.6(a)*) uses a single phototransistor output stage and is usually housed in a 6-pin package, with the base terminal of the phototransistor externally available. In normal use the base is left open circuit, and under this condition the optocoupler has a minimum CTR value of 20% and a useful bandwidth of 300kHz. The phototransistor can, however, be converted to a photodiode by shorting the base (pin-6) and

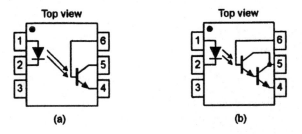

Figure 7.6 *Typical simple (a) and Darlington (b) isolating optocouplers*

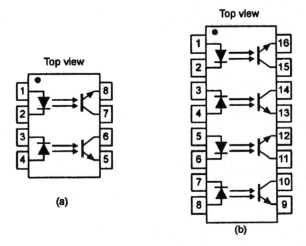

Figure 7.7 *Typical dual (a) and quad (b) isolating optocouplers*

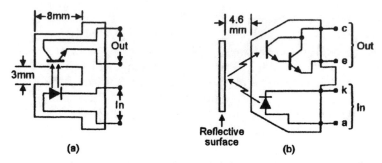

Figure 7.8   *Typical slotted (a) and reflective (b) optocouplers*

Parameter	Isolating optocouplers				Slotted optocoupler	Reflective optocoupler
	Simple type	Darlington type	Dual type	Quad type		
Isolating voltage	±4kV	±4kV	±1.5kV	±1.5kV	N.A.	N.A.
V_{CE} (max)	30V	30V	30V	30V	30V	15V
I_F (max)	60mA	60mA	100mA	100mA	50mA	40mA
CTR (min)	20%	300%	12.5%	12.5%	10%	0.5%
Bandwidth	300kHz	30kHz	200kHz	200kHz	300kHz	20kHz
Outline	Fig 7.6(a)	Fig 7.6(b)	Fig 7.7(a)	Fig 7.7(b)	Fig 7.8(a)	Fig 7.8(b)

Figure 7.9   *Typical parameter values of the Figure 7.6 to 7.8 devices*

emitter (pin-4) terminals together; under this condition the CTR value falls to about 0.2% but the bandwidth rises to about 30MHz.

The Darlington optocoupler (*Figure 7.6(b)*) is also housed in a 6-pin package and has its phototransistor base externally available. Because of the high current gain of the Darlington, this coupler has a typical minimum CTR value of about 300%, but has a useful bandwidth of only 30kHz.

The dual and quad optocouplers of *Figure 7.7* use single-transistor output stages in which the base terminal is not externally available.

Note in all four isolating devices that the input pins are on one side of the package and the output pins on the other. This construction gives the maximum possible values of isolating voltage. Also note in the multichannel devices of *Figure 7.7* that, although these devices have isolating voltages of 1.5kV, potentials greater than 500V should not be allowed to exist between adjacent channels.

Isolating voltage values are not specified for the slotted and reflective optocoupler devices of *Figure 7.8*. The *Figure 7.8(a)* device has a typical slot width of about 3mm, and uses a single output transistor to give an open slot minimum CTR value of 10% and a bandwidth of 300kHz.

Finally, the reflective optocoupler of *Figure 7.8(b)* uses a Darlington output stage and has a useful bandwidth of only 20kHz. Even so, the device has a typical minimum CTR value of only 0.5% at a reflective range of 5mm from a surface with a reflective efficiency of 90%, when the input LED is operated at its maximum current of 40mA.

Optocoupler usage notes

Optocouplers are very easy devices to use, with the input side being used in the manner of a normal LED and the output used in the manner of a normal phototransistor, as described in earlier chapters. The following notes give a summary of the salient usage points.

The input current to the optocoupler LED must be limited via a series-connected external resistor which, as shown in *Figure 7.10*, can be connected on either the anode or the cathode side of the LED. If the LED is to be driven from an AC source, or there is a possibility of a reverse voltage being applied across the LED, the LED must be protected from reverse voltages via an external diode connected as shown in *Figure 7.11*.

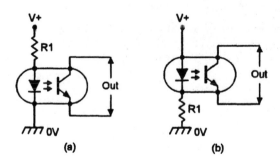

Figure 7.10 *The LED current must be limited by a series resistor, which can be connected to either the anode (a) or the cathode (b)*

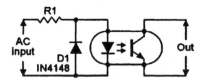

Figure 7.11 *The input LED can be protected against reverse voltages via an external diode*

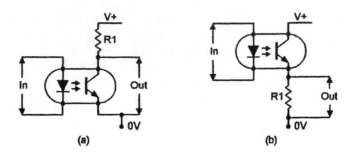

Figure 7.12   *An external output resistor, wired in series with the phototransistor, can be connected to either the collector (a) or emitter (b)*

The phototransistor's operating current can be converted into a voltage by wiring an external resistor in series with the collector of the device. This resistor can be connected to either the collector or the emitter of the photo-transistor, as shown in *Figure 7.12*. The greater the value of this resistor, the greater is the sensitivity of the circuit but the lower is its bandwidth.

In normal use, the phototransistor is used with its base terminal open-circuit. If desired, however, the phototransistor can be converted into a photodiode by using the base terminal as shown in *Figure 7.13(a)* and ignoring the emitter terminal (or shorting it to the base). This connection results in a greatly increased bandwidth (typically 30MHz), but a greatly reduced CTR value (typically 0.2%).

Alternatively, the base terminal can be used to vary the CTR value of the optocoupler by wiring an external resistor (RV1) between the base and emitter, as shown in the Darlington example of *Figure 7.13(b)*. With RV1 open-circuit, the CTR value is that of a normal Darlington optocoupler (typically 300% minimum); with RV1 short-circuit, the CTR value is that of a diode-connected phototransistor (typically about 0.2%).

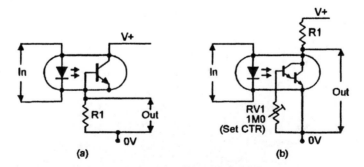

Figure 7.13   *If its base is available, the phototransistor can be made to function as a photodiode (a), or its CTR values can be varied via RV1 (b)*

Digital interfacing

Optocoupler devices are ideally suited for use in digital interfacing applications in which the input and output circuits are driven by different power supplies. They can be used to interface digital ICs of the same family (TTL, CMOS, etc.) or digital ICs of different families, or to interface the digital outputs of home computers, etc., to motors, relays and lamps, etc. This interfacing can be achieved using various special-purpose 'digital interfacing' optocoupler devices, or by using standard optocouplers; *Figures 7.14 to 7.16* show circuits of the latter type.

Figure 7.14 shows how to interface two TTL circuits, using an optocoupler circuit that provides a non-inverting action. Here, the optocoupler LED and current-limiting resistor R1 are connected between the 5V positive supply rail and the output-driving terminal of the TTL device (rather than between the TTL output and ground), because TTL outputs can usually sink a fairly high current (typically 16mA) but can source only a very low current (typically 400µA).

The open-circuit output voltage of a TTL IC falls to less than 0.4V when in the logic-0 state, but may rise to only 2.4V in the logic-1 state if the IC is not fitted with an internal pull-up resistor. In such a case the optocoupler LED current will not fall to zero when the TTL output is at logic-1. This snag is overcome in the *Figure 7.14* circuit by fitting an external pull-up resistor (R3) as shown. The *Figure 7.14* circuit's optocoupler phototransistor is wired between the input and ground of the driven (right-hand) TTL IC because a TTL input needs to be pulled down to below 800mV at 1.6mA to ensure correct logic-0 operation.

CMOS IC outputs can source or sink currents (up to several mA) with equal ease. Consequently, these devices can be interfaced by using a sink configuration similar to that of *Figure 7.14*, or they can use the source configuration shown in *Figure 7.15*. In either case, the R2 value must be large enough to provide an output voltage swing that switches fully between the CMOS logic-0 and logic-1 states.

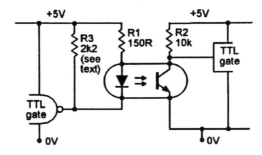

Figure 7.14 *TTL interface*

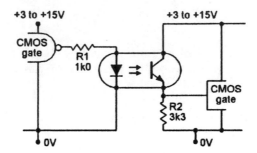

Figure 7.15 *CMOS interface*

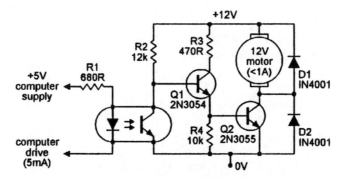

Figure 7.16 *Computer-driven DC motor*

Figure 7.16 shows how the optocoupler can be used to interface a computer's output signal (5V, 5mA) to a 12V DC motor that draws an operating current of less than 1A. With the computer output high, the optocoupler LED and phototransistor are both off, so the motor is driven on via Q1 and Q2. When the computer output goes low, the LED and phototransistor are driven on, so Q1–Q2 and the motor are cut off. The reverse of this action can be obtained by wiring the optocoupler's output in series between R2 and Q1-base, so that Q1–Q2 and the motor turn on only when the computer output goes low.

Analogue interfacing

An optocoupler can be used to interface analogue signals from one circuit to another by setting up a standing current through the LED and then modulating this current with the analogue signal. *Figure 7.17* shows this technique used to make an audio-coupling circuit.

Here, the op-amp is connected in the unity-gain voltage follower mode, with the optocoupler LED wired into its negative feedback loop so that the

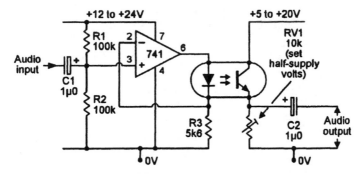

Figure 7.17 *Audio-coupling circuit*

voltage across R3 (and thus the current through the LED) precisely follows the voltage applied to the op-amp's pin-3 non-inverting input terminal. This terminal is dc biased at half-supply volts via the R1–R2 potential divider, and can be ac-modulated by an audio signal applied via C1. The quiescent LED current is set at 1 to 2mA via R3.

On the output side of the optocoupler, a quiescent current is set up (by the optocoupler action) in the phototransistor, and causes a quiescent voltage to be set up across RV1, which should have its value adjusted to give a quiescent output value of half-supply voltage. The audio output signal appears across RV1 and is dc-decoupled via C2.

Triac interfacing

An ideal application for the optocoupler is that of interfacing the output of a low-voltage control circuit (possible with one side of its power supply

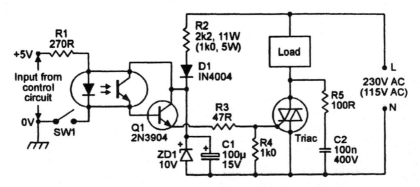

Figure 7.18 *Simple non-synchronous triac power switch with optocoupled input*

grounded) to the input of a triac power-control circuit that is driven from the AC power lines and which can be used to control the power feed to lamps, heaters, and motors. *Figure 7.18* shows an example of such a circuit; the figures in parenthesis show the component values that should be used if 115V AC (rather than 230V) supplies are used; the actual triac type must be chosen to suit individual load/supply requirements.

The *Figure 7.18* circuit gives a non-synchronous switching action in which the triac's initial switch-on point is not synchronized to the AC power line waveform. Here, R2–D1–ZD1 and C1 are used to develop an AC-derived 10V DC supply, which can be fed to the triac gate via Q1 and hence used to turn the triac on and off. Thus, when SW1 is open the optocoupler is off, so zero base drive is applied to Q1, and the triac and load are off. When SW1 is closed, the optocoupler drives Q1 on and connects the 10V DC supply to the triac gate via R3, thus applying full AC mains power to the load.

Optocoupled SCRs and triacs

SCRs (silicon controlled rectifiers) and triacs are semiconductor power-switching devices that (like transistors) are inherently photosensitive. An optocoupled SCR is simply an SCR and a LED mounted in a single package, and an optocoupled triac is simply a triac and a LED mounted in a single package. Such devices are readily available, in both simple and complex forms; some sophisticated triac types incorporate interference-suppressing zero-crossing switching circuitry in the package.

Figures 7.19(a) and *7.19(b)* show the typical outlines of simple optocoupled SCRs and triacs (which are usually mounted in 6-pin DIL packages); *Figure 7.20* lists the typical parameters of these two particular devices, which have rather limited rms output-current ratings, the values being (in the examples shown) 300mA for the SCR and 100mA for the triac. The SCR device's surge-current rating is 5A at a pulse width of 100µs and a duty cycle of less than 1%; the triac device's surge rating is 1.2A at a pulse width of 10µs and a duty cycle of 10% maximum.

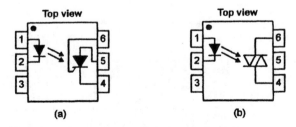

(a)   **(b)**

Figure 7.19   *Typical optocoupled SCR (a) and triac (b)*

Parameter	Optocoupled SCR	Optocoupled triac
LED characteristic I_F (max)	60mA	50mA
SCR/triac characteristic V_MAX	400V	400V
I_MAX (rms)	300mA	100mA
I_SURGE (see text)	5A	1.2A
Coupling characteristic Isolating voltage	±1.5kV	±1.5kV
Input trigger current	5mA typical (20mA max)	5mA typical (20mA max)

Figure 7.20   *Typical characteristics of optocoupled SCRs/triacs*

Optocoupled SCRs and triacs are very easy to use; the input LED is driven in the way of a normal LED, and the SCR/triac is used like a normal low-power SCR/triac. *Figures 7.21* to *7.23* show various ways of using an opto-coupled triac; R1 should be chosen to pass a LED current of at least 20mA; all other component values are those used with a 230V AC supply.

In *Figure 7.21* the triac is used to directly activate an AC line-powered filament lamp, which should have an rms rating of less than 100mA and a peak inrush current rating of less than 1.2A.

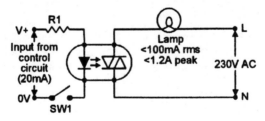

Figure 7.21   *Low-power lamp control*

Figure 7.22 shows how the optocoupled triac can be used to activate a slave triac, and thereby activate a load of any desired power rating. This circuit is suitable for use only with non-inductive loads such as lamps and heating elements, using a triac of suitable rating.

Finally, *Figure 7.23* shows how the above circuit can be modified for use with inductive loads such as electric motors. The R2–C1–R3 network provides a degree of phase-shift to the triac gate-drive network, to ensure correct triac triggering action, and R4–C2 form a snubber network, to suppress rate-of-rise (rate) effects.

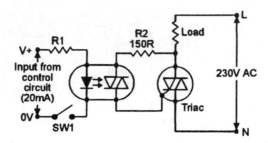

Figure 7.22   *High-power control via a triac slave*

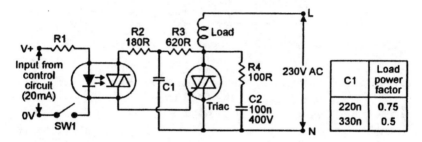

Figure 7.23   *Driving an inductive load*

Optocoupled SSRs

An optocoupled solid-state relay (SSR) is a device that can be used as a superior replacement for many types of low-power electromechanical relay. Like a normal relay, it provides complete electrical isolation between its input and output circuits, and its output acts like an electrical switch that has a near-infinite resistance when open and a very low resistance when closed and which, when closed, can pass AC or DC currents with equal ease, without suffering significant 'offset voltage' losses.

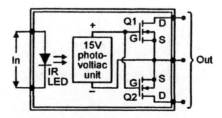

Figure 7.24   *Basic Siemens SSR design*

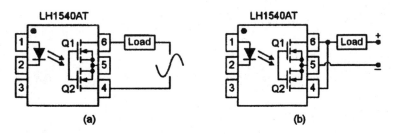

Figure 7.25 *Method of connecting the LH1540AT's outputs for use with loads driven by (a) AC or (b) DC supplies*

Siemens are the present market leaders in the optocoupled SSR field. *Figure 7.24* illustrates their basic SSR design, which has an IR LED input stage and a dual *n*-channel enhancement-type MOSFET output stage that (unlike a dual bipolar transistor stage) does not produce significant offset voltage drops when biased on. The IR LED's output is coupled to the inputs of the MOSFETs via a bank of 25 photovoltaic diodes that, when illuminated, apply a 15V turn-on voltage to the MOSFET gates.

The simplest device in the Siemens range of optocoupled SSRs is the LH1540AT, which is housed in a 6-pin package and has an output that acts as a normally open (NO) single-pole switch. The device has an isolation voltage rating of 3.75kV and a maximum output load voltage rating of 350V. *Figure 7.25* shows basic details of the LH1540AT and shows how to connect its output pins for use with loads driven by (a) AC or (b) DC supplies; the two output IGFETs are effectively connected in inverse series for AC operation, or in parallel for DC operation. When the input LED is passing a current of 5mA, the output can handle maximum load currents of 120mA and has a typical 'on' resistance of 25 ohms when used in the AC configuration, or 250mA and 5 ohms in the DC configuration. The device has typical on/off switching speeds of less than 1ms.

Other devices in the Siemens optocoupled SSRs range include ones that have outputs that act as singe-pole or 2–pole NC, NO, or change-over switches.

8

Brightness-control circuits

In most applications, simple light-generating devices such as filament lamps and LEDs are used in the basic ON/OFF mode, in which they are always either at full brilliance or are fully off. In fact, however, the brightness of these devices is fully variable between these two extremes; this chapter presents a variety of practical brightness-control circuits.

Lamp-control basics

Ordinary filament lamps can be powered from AC or DC supplies and consist of a coil of tungsten wire (the filament) suspended within a vacuum-filled glass envelope and connected to the outside world via a pair of metal terminals; the filament runs white hot when connected to a suitable external voltage, thus generating a bright white light.

In this type of lamp, the filament's resistance has a positive temperature coefficient, causing the lamp resistance to increase with operating temperature; *Figure 8.1* shows the typical resistance variations that occur in a 12V

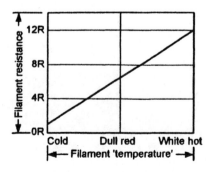

Figure 8.1 *Graph showing variations in filament resistance with filament temperature for a 12V, 12W lamp*

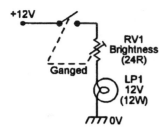

Figure 8.2 *Rheostat brightness-control circuit*

12W lamp. Thus, the resistance is 12R when the filament is operating at its normal 'white' heat, but only 1R0 when it is cold. This 12:1 resistance variation is typical of all tungsten filament lamps and causes them to have switch-on inrush current values about twelve times greater than the normal running values.

The brilliance of a DC-powered filament lamp can be varied in any one of three basic ways. The simplest way is to wire a rheostat and a ganged switch in series with the lamp, as in *Figure 8.2*, which shows a 12V circuit that drives a 12W lamp. Here, if RV1 has a maximum resistance value double that of the 'hot' resistance value of the lamp, RV1 will enable the lamp power dissipation (and thus its brilliance) to be varied over an approximately 20:1 range, as explained below.

The operation of the *Figure 8.2* rheostat circuit is fairly simple, as shown in *Figure 8.3*. When RV1 is set to the maximum brilliance (zero RV1 resistance) position, the full 12V supply is applied to the lamp, which presents a resistance of 12 ohms and thus has a power consumption of 12W, as shown in *Figure 8.3(a)*. When RV1 is set to the minimum brilliance (maximum RV1 resistance) position, however, RV1 has a resistance of 24 ohms and the lamp presents a resistance of only 3R0, as shown in *Figure 8.3(b)*, and under this condition only 1.33V is developed across the lamp, which thus consumes only 590mW and produces very little light output. RV1 thus allows the lamp brilliance to be varied over a wide range.

A major disadvantage of the *Figure 8.2* circuit is that it wastes a lot of power in RV1, which must have a substantial power rating and be capable

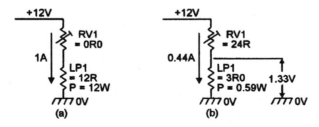

Figure 8.3 *Equivalents of the Figure 8.2 circuit at (a) maximum and (b) minimum brightness levels*

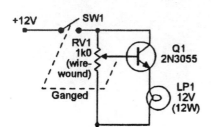

Figure 8.4   *Variable-voltage brightness-control circuit*

of handling the lamp's 'cold' current. *Figure 8.4* shows an alternative brilliance-control circuit, which dissipates negligible power in RV1. Here, RV1 acts as a variable potential divider which applies an input voltage to the base of emitter follower Q1, which buffers (power boosts) this voltage and applies it to the lamp. RV1 thus enables the lamp voltage (and brilliance) to be fully varied from zero to maximum. Note in this circuit that Q1 needs a fairly large power rating and must be capable of handling the cold currents of the lamp.

Switched-mode control

The third and most sophisticated way of controlling the brilliance of a DC-powered lamp is the so-called 'switched-mode' method, which is shown in basic form in *Figure 8.5*. Here, an electronic switch (SW1) is wired in series with the lamp and can be opened and closed via a pulse-generator waveform. When this pulse is high, SW1 is closed and power is fed to the lamp; when the pulse is low SW1 is open and zero power is fed to the lamp.

The important thing to note about the *Figure 8.5* pulse generator is that it generates a waveform with a fixed frame width but with a variable mark/space (ON/OFF) ratio or duty cycle, thereby enabling the MEAN lamp voltage to be varied. Typically, the M/S ratio is fully variable from 1:20 to

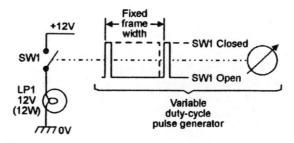

Figure 8.5   *Basic switched-mode brightness-control circuit*

20:1, enabling the mean lamp voltage to be varied from 5% to 95% of the supply-voltage value.

Because of the inherently long thermal time constant of a tungsten filament lamp, its brilliance responds relatively slowly to rapid changes in input power. Consequently, if the frame width of the *Figure 8.5* waveform generator is less than roughly 25ms (i.e. the repitition frequency is greater than 40Hz), the lamp will show no sign of flicker, and the lamp brilliance can be varied by altering the M/S ratio.

Thus, if the M/S ratio of the *Figure 8.5* circuit is set at 20:1, the mean lamp voltage is 11.4V and the consequently 'hot' lamp consumes 10.83W. Alternatively, with the M/S ratio set at 1:20, the mean lamp voltage is only 600mV, so the lamp is virtually cold and consumes a mere 120mW. The lamp power consumption can thus be varied over a 90:1 range via the M/S-ratio control. Note, however, that this wide range of power control is obtained with very little power loss within the system, since power is actually controlled by SW1, which is always either fully on or fully off. The switched-mode control system is thus highly efficient.

Figure 8.6 shows the practical circuit of a switched-mode DC lamp dimmer or brilliance control that is designed for use with a 12V lamp with a maximum power rating of 24W and enables the lamp's light intensity to be smoothly varied from zero to full brilliance via 100k variable resistor RV1. The circuit operates as follows.

In *Figure 8.6*, IC1a and IC1b (comprising one half of a 4001B CMOS quad 2-input NOR gate) are wired as an astable multivibrator or squarewave generator in which half of the waveform is controlled via C1–D1–R1 and the right-hand part of RV1, and the other half is controlled via C1–D2–R2 and the left-hand part of RV1, thus enabling the M/S-ratio to be varied via RV1.

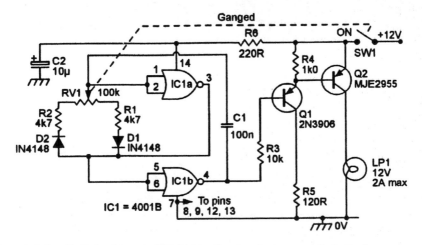

Figure 8.6 *Switched-mode DC lamp dimmer (–ve ground version)*

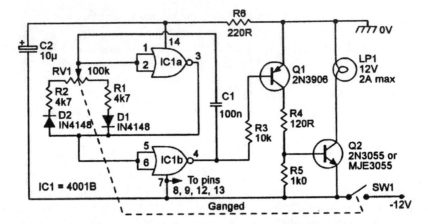

Figure 8.7   *Switched-mode DC lamp dimmer (+ve ground version)*

Thus, when SW1 is closed the astable operates and feeds a switching waveform to the lamp via Q1 and Q2. The astable operates at a fixed frequency of about 100Hz, but its M/S-ratio is fully variable from 1:20 to 20:1 via RV1, thus enabling the mean lamp power to be varied over a 90:1 range. Note that ON/OFF switch SW1 is ganged to RV1, so that the circuit can be switched fully off by turning the RV1 brilliance control fully anticlockwise.

The *Figure 8.6* lamp dimmer circuit can be used to control the brilliance of virtually any low power (up to 24 watts) filament lamps that are powered by 12 volt DC supplies. Note, however, that if it is used to control vehicle lights it can only be used in ones in which the 'free' ends of the lamps go to the +ve supply line via control switches, as is normal in vehicles fitted with negative-ground electrical systems (in which the −ve battery terminal goes to the chassis of the vehicle).

The alternative circuit of *Figure 8.7* can be used to control the lights of vehicles in which the free end of the lamp goes to the -ve supply line via control switches, as occurs in some old vehicles fitted with positive-ground electrical systems. Note in both the *Figure 8.6* and *8.7* circuits that R6–C2 are used to protect IC1 against damage from high-voltage transients that may occur on the vehicle's supply lines.

LED brightness control

The brilliance of LEDs and 7-segment LED displays can be controlled by varying the device's DC operating current, or by using the switched-mode technique to vary the mean voltage applied to the device via its current-limiting resistor(s). If the latter option is used, however, it must be noted that these devices give an instant response to changes in input power level, so the

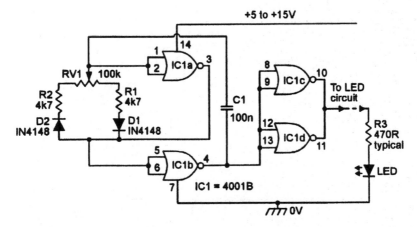

Figure 8.8 *Switched-mode LED brightness-control circuit*

design technique must rely on the natural integrating action of the human eye to ensure a flicker-free brightness-control action.

The natural action of the human eye is such that it ignores the instantaneous values of rapidly repeating changes in light level if these changes occur at a frequency in excess of about 40Hz, and sees these changes in terms of the mean value of light intensity instead. Thus, in LED brightness-control circuits, the variable M/S-ratio generator normally operates in the range 50Hz to 100Hz.

Figure 8.8 shows a practical example of a LED brightness-control circuit, designed around a single 4001B CMOS IC. Here, IC1a–IC1b are wired as a 100Hz astable, with M/S-ratio variable from 1:20 to 20:1 via RV1, and IC1c and IC1d are connected in parallel to provide a medium-current (15 to 20mA) buffered drive to the LED via current-limiter R3.

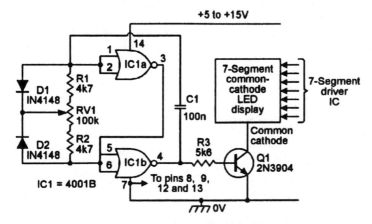

Figure 8.9 *Common-cathode 7-segment LED brightness-control circuit*

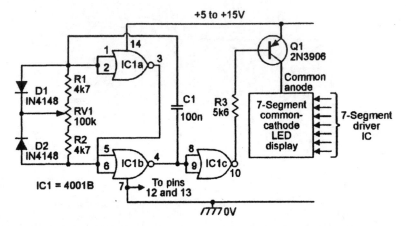

Figure 8.10   *Common-anode 7-segment LED brightness-control circuit*

Figure 8.9 shows how to apply switched-mode brightness control to a common-cathode 7-segment LED display. Here, IC1a–IC1b are again wired as a 100Hz astable with variable M/S-ratio, but in this case the output waveform is fed (via R3) to the base of Q1, which functions as a medium-power switch that is wired in series with the common cathode terminal of the display.

Figure 8.10 shows how to modify the above circuit for use with a common-anode display. Here, IC1c is used as an inverting buffer that connects the astable output signal to the base of *pnp* transistor Q1.

In practice, many 7-segment LED driver ICs have a 'blanking' terminal (enabling the display to be turned on and off) which can be used to apply switched-mode brightness control to the LED display device. This terminal

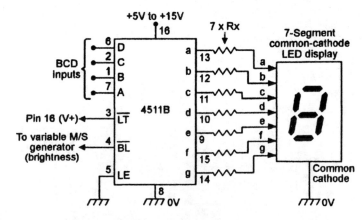

Figure 8.11   *7-segment LED brightness-control via a 4511B driver IC*

is usually designated BL or not-BL. The old TTL 74LS47 and 74LS48 range of decoder/driver ICs have such a terminal, as also does the CMOS 4511B latch/decoder/driver. In the latter case, the display is blanked when the not-BL pin is low, and is active when the pin is high. *Figure 8.11* shows how to connect this IC to give brightness control via an external variable M/S-ratio generator (such as is used in *Figures 8.6* to *8.10*).

AC lamp-control basics

The brilliance of an AC-powered lamp can be controlled by using a triac and a variable phase-delay network to vary the power feed to the lamp, as shown in the basic phase-triggered system of *Figure 8.12*. The triac is a bidirectional (AC) solid-state self-latching power switch that can be turned on by applying a brief trigger pulse to its gate, but which turns off again automatically at the end of each power half-cycle as its main-terminal currents fall to near-zero.

Thus, in *Figure 8.12*, the triac is triggered via a variable phase-delay network that is interposed between the AC power line and the triac gate. Hence, if the triac is triggered 10° after the start of each half-cycle, almost the full available r.m.s supply voltage is fed to the lamp load. If the triac is triggered 90° after the start of each half cycle, only half of the r.m.s. line voltage is fed to the load. Finally, if the triac is triggered 170° after the start of each half-cycle (e.g. 10° before the end of each half-cycle), only a very small part of the available r.m.s. line voltage is fed to the load.

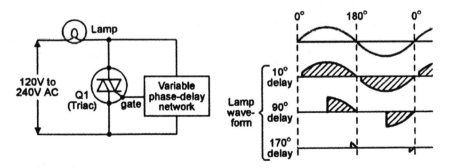

Figure 8.12 *Basic phase-triggered triac AC brightness-control circuit and waveforms*

The three most popular methods of obtaining variable phase-delay triggering are to use either a line-synchronized UJT (*UniJunction Transistor*), or a special-purpose IC, or to use a diac and an R–C network in the basic configuration shown in *Figure 8.13*. The diac is a bilateral threshold switch which,

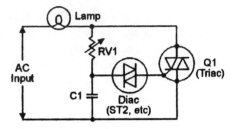

Figure 8.13   *Basic diac-type variable phase-delay lamp dimmer circuit*

when connected across a voltage source, presents a high impedance until the applied voltage rises to about 35 volts, at which point the device switches into a low-impedance state and remains there until the applied voltage falls to about 30 volts, at which point it reverts back to the high-impedance mode.

Thus, in *Figure 8.13*, in each AC power-line half-cycle the RV1–C1 network applies a variable phase-delayed version of the power-line waveform to the triac gate via the diac, and each time that the C1 voltage rises to 35 volts the diac fires and delivers a trigger pulse to the triac gate, thus turning the triac on and simultaneously applying power to the lamp load and removing the drive from the RV1–C1 network. The mean power to the load (integrated over a full half-cycle period) is thus fully variable from near-zero to maximum via RV1.

Practical design considerations

The basic phase-delay lamp dimmer circuit of *Figure 8.13* operates at potentially lethal voltages and must be modified in two basic ways before it can be used as a safe and 'legal' design. The first modification is concerned purely with the matter of safety, as follows.

All normal AC-powered lighting systems operate at potentially lethal voltages; their design is thus strictly controlled by legal regulations. *Figure 8.14(a)* shows a legal domestic lighting circuit. It is powered via the AC supply's neutral (N) and live (L, or 'hot') lines; one of the lamp socket's two terminals is wired directly to the neutral line, and the other is connected to the live line via isolating switch SW1; this design ensures the lamp and its socket are completely isolated from the live line when SW1 is open and thus present negligible electrical hazard. Note that it is now quite illegal to place the lamp and its socket on the live side of SW1.

Simple '2-wire' lamp dimmers of the *Figure 8.13* type are designed to be wired in series with the lamp, and are thus known as *series* dimmers; *Figure 8.14(b)* shows how such a dimmer must be connected into the lamp's circuit to comply with legal safety requirements. Some dimmers are '3-wire' types,

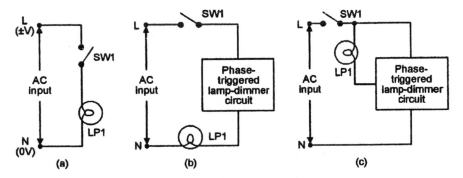

Figure 8.14 *Ways of controlling an AC-powered lamp via (a) a simple switch, (b) a switch + series dimmer, and (c) a switch + paralleled series dimmer*

in which the dimmer circuit must (when SW1 is closed) be connected in parallel with the live and neutral terminals, but the lamp must be wired in series with SW1 and the dimmer's third wire; dimmers of this type are known as *paralleled series* dimmers; *Figure 8.14(c)* shows how such a dimmer must be connected into the lamp's circuit to comply with legal safety requirements.

The second modification that must be made to the basic phase-delay lamp dimmer circuit of *Figure 8.13* to ensure that it complies with legal requirements is concerned with the matter of RFI generation, and can be understood with the help of *Figure 8.12*. Note from the *Figure 8.12* waveforms that each time the triac is gated on, the load current transitions sharply (in a few microseconds) from zero to a value determined by the lamp resistance and the value of instantaneous AC power-line voltage. These transitions inevitably generate RFI (radio-frequency interference); the RFI is greatest when the triac is triggered at 90°, and is least when the triac is triggered close to the 0° and 180° 'zero crossing' points of the power-line waveform.

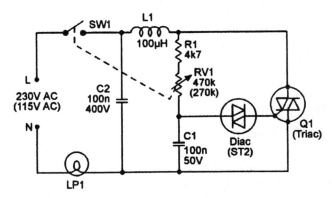

Figure 8.15 *Practical circuit of a simple lamp dimmer with RFI suppression*

In lamp-dimmer brightness-control circuits, where there may be considerable lengths of power cable between the triac and the lamp load, this RFI can be offensive. Consequently, in practical lamp dimmers, the circuit is usually provided with an L–C RFI-suppression network, as shown in the circuit of *Figure 8.15*; this circuit also shows how ON/OFF switch SW1 can be ganged to brightness control pot RV1.

Note in *Figure 8.15* (and all other triac circuits shown in this chapter) that the diac can be virtually any commercial type (ST2, etc), that the triac type should be chosen to suit the power-line voltage and lamp-load values, and that the values shown in brackets are applicable to 115V (rather than 230V) power-line operation.

Backlash reduction

The simple *Figure 8.15* circuit makes a useful lamp dimmer but has one annoying characteristic, in that RV1 has considerable hysteresis or backlash. If, for example, the lamp finally goes fully off when RV1 is increased to 470k (in the 230V circuit), it may not start to go on again until RV1 is reduced to 400k, and it then suddenly burns at a fairly high brightness level. The cause of this characteristic is as follows:

The basic action of the *Figure 8.15* circuit is such that, in the first part of each power-line half-cycle, C1 charges via RV1–R1, etc., until C1 charges to 35 volts, at which point the diac suddenly fires and starts to partially discharge C1 into the gate of the triac; as the triac turns on it connects the remaining part of the half-cycle to the lamp, thus removing the power-line drive from R1–RV1. This switching action only takes 2μs or so, but in this brief period the diac is able (in the *Figure 8.15* circuit) to remove substantial charge (typically about 5V) from C1 and thus upsets the timing of the following half-cycle, thereby causing the annoying backlash characteristic.

One easy way to reduce this backlash is to simply wire a current-limiting resistor (47R to 120R) in series with the diac, to reduce the amount of C1 voltage change that takes place in the 2μs triac-switch-on period. Another way is to use the gate-slaving technique shown in *Figure 8.16*. This circuit is similar to that of *Figure 8.15*, except that the charge of C1 is coupled to slave capacitor C2 via the relatively high resistance of R2. C1 thus charges to a slightly higher voltage than C2, and C2 fires the diac once its voltage reaches 35V. Once the diac has fired it reduces the C2 potential briefly to 30V, but has little influence on the voltage value of the C1 main-timing capacitor; the circuit backlash is thus reduced. The backlash can be reduced even more by wiring a current-limiting resistor in series with the diac (as described above), to reduce the magnitude of the C2 discharge voltage, as shown in *Figure 8.17*.

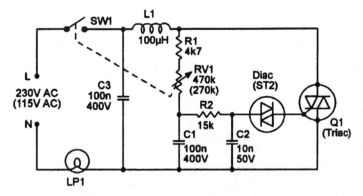

Figure 8.16 *Improved lamp dimmer with gate slaving*

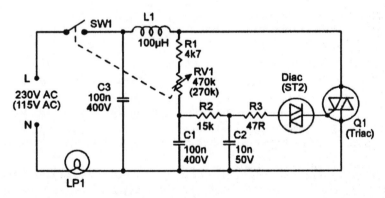

Figure 8.17 *Minimum-backlash lamp dimmer*

UJT triggering

A lamp dimmer with absolutely zero backlash can be made by using a power-line-synchronized variable-delay unijunction transistor (UJT) circuit to trigger the lamp-driving triac in each AC half-cycle. *Figure 8.18* shows such a circuit, which acts as a paralleled series dimmer (see *Figure 8.14(c)*). Here, the UJT is powered from a 12V DC supply derived from the AC power line via R1–D1–ZD1 and C1. The UJT is synchronized to the AC power-line via the Q2–Q3–Q4 zero-crossing detector network, the action being such that Q4 is turned on (applying power to the UJT circuit) via Q2 or Q3 at all times other than when the instantaneous power-line voltage is close to the zero-crossover point at the end or the start of each power-line half-cycle.

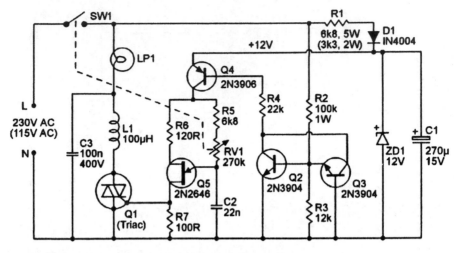

Figure 8.18   *UJT-triggered zero-backlash lamp dimmer*

Thus, shortly after the start of each half-cycle, power is applied to the UJT circuit via Q4, initiating the start of the UJT timing cycle, and a short time later (determined by R5–RV1–C2) the UJT (Q5) delivers a trigger pulse to the triac gate, driving the triac on and connecting power to the lamp load for the remaining part of the half-cycle. The triac and the UJT circuit automatically reset at the end of each half-cycle, and a new sequence then begins.

The *Figure 8.18* circuit generates absolutely zero control backlash, and can be usefully modified for use in a variety of non-standard applications. *Figure 8.19*, for example, shows a circuit that can be fitted in place of the existing UJT network to modify the *Figure 8.18* circuit so that it acts as a slow-start

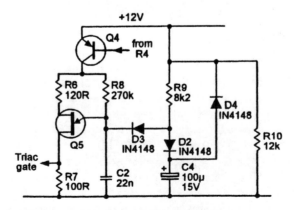

Figure 8.19   *Slow-start lamp-control circuit (for use with Figure 8.18)*

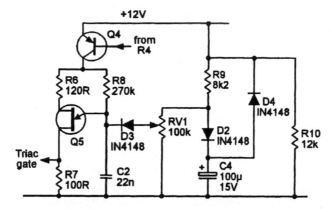

Figure 8.20 *Combined lamp dimmer and slow-start circuit (for use with Figure 8.18)*

lamp dimmer that simply causes the lamp brilliance to rise slowly from zero to maximum when first turned on, taking about two seconds to reach full brilliance. The circuit is intended to eliminate high turn-on inrush currents and to thus extend the lamp life. Circuit operation is as follows:

When power is first applied to the circuit, C4 is fully discharged and acts like a short-circuit, so C2 charges via high-value resistor R8 only; the UJT thus generates a long delay under this condition, so the triac is triggered late in each half-cycle and the lamp burns at low brilliance. As time passes, C4 slowly charges up via R9, enabling the C2 charge to be supplemented via R9–D3, thereby progressively reducing the UJT time constant and increasing the lamp brilliance until, when C4 is fully charged (after roughly two seconds), full brilliance is reached.

Figure 8.20 shows how the above circuit can be further modified so that it acts as a combined lamp dimmer (via RV1) and a slow-start circuit. Note in these two circuits that diode D2 prevents C4 from discharging into the UJT each time Q5 fires, and D4 automatically discharges C4 via R10 and thus resets the network when the circuit is turned off.

IC-controlled lamp dimmer circuits

Many modern lamp dimmers have their triac driven via a dedicated 'smart' IC that can turn the lamp on or off or control its brilliance, the IC taking its action commands via a touch-sensitive pad or push-button input switch, etc. Siemens are the leading producers of this type of IC; their first really successful IC was the S566B, which incorporated touch-conditioning circuitry. Its action was such that a very brief input touch or push made the lamp change state (from OFF to a remembered ON state, or vice versa), but a sustained

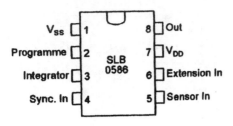

Figure 8.21   *SLB0586 outline and pin notations*

(greater than 400ms) 'dimming' input touch or push made the IC go into a ramping mode in which the lamp power slowly ramped up and down (between 3 and 97 percent of maximum) until the input was released, at which point the prevailing power level was held and remembered; the ramp direction reversed on alternate dimming touches.

Over the years, the S566B has been replaced by a succession of improved Siemens designs, their current product (introduced in 1990) being the SLB0586 (see *Figure 8.21*), which draws a mean current of 0.45mA from a 5.6V DC supply and allows the user to select any of three alternative dimming modes via the pin-2 *Programme* input terminal. *Figure 8.22* shows the IC's basic application circuit, using a single touch-sensitive control input (made from a simple 'button' pad or strip of conductive material).

Note in *Figure 8.22* that the lamp is connected directly to the neutral AC supply line and is connected to the live line via the series-type dimmer circuit and a fuse. The IC is powered (between pins 1 and 7) from a 5V6 DC supply derived from the AC line via R2–C2–ZD1–D1–C3, and has its *Extension In* input disabled by shorting pin 6 to pin 7. Its pin-5 *Sensor In* input works on the inductive pick-up principle, in which the human body picks up radiated

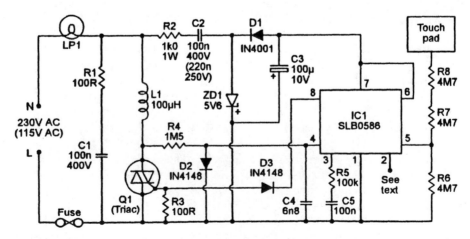

Figure 8.22   *Basic SLB0586 lamp dimmer circuit, with touch-sensitive control*

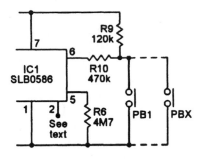

Figure 8.23 *Push-button control applied to the Figure 8.22 circuit*

AC power-line signals, which the IC detects when the touch pad is touched; the pad must be placed close to the IC, to avoid unwanted pick-up. The operator is safely protected from the AC power-line voltage via R7–R8; for correct operation the AC supply *must* be connected as shown, with the live or 'hot' lead to pin-1 of the IC, and the neutral line to the lamp.

The *Figure 8.22* lamp dimmer circuit can be set to any one of three basic operating modes by suitably using the pin-2 *Programme* terminal. If pin-2 is left open, the circuit gives the same ON/OFF and ramping-control action as already described for the S566B. If, on the other hand, pin-2 is shorted to pin-7, the lamp always goes to maximum brightness when switched ON, and in dimming operations the lamp starts at minimum brightness and then slowly ramps up and down until the sensor is released; the ramping direction does not reverse on successive dimming operations. Finally, if pin-2 is shorted to pin-1, the lamp operation is like that just described except that the ramping direction reverses on alternate dimming operations.

The basic *Figure 8.22* circuit can be modified in various ways. If, for example, multi-input operation is required, it can be obtained by wiring any desired number of push-button control switches in parallel and modifying the circuit as shown in *Figure 8.23*. Here, the pin-6 to pin-7 connection is broken and replaced by the R9–R10 divider. The push-button control switches are connected between the R9–R10 junction and pin-1 of the IC. If the 'touch

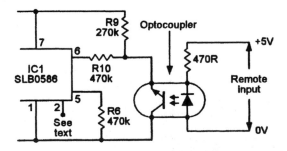

Figure 8.24 *Optocoupler operation of the SLB0586 lamp dimmer circuit*

control' facility is not needed, R7–R8 can be eliminated and R6 can be reduced to 470k, as shown.

Finally, *Figure 8.24* shows how to modify the lamp dimmer so that it can be controlled by an external circuit via an optocoupler, without the use of the touch control facility.

Light, mirrors, prisms and lenses

So far, this book has dealt mainly with the electronics side of optoelectronics and has made very little mention of the purely optical ('opto') aspects of the subject. This chapter aims to help remedy this situation by giving fairly concise descriptions of vital optoelectronics-related subjects such as the nature, behaviour, and applications of light, mirrors, prisms and lenses. Chapter 10 continues this theme by describing the basic principles of fibre optics and of lasers.

Light

Light is a form of energy and is transported by electromagnetic radiation. It has an apparent dualistic nature that enables it to be regarded as both a wave phenomenon and as a flux-like flow of sub-atomic particles known as *photons*, which are released as a consequence of shifts in the energy levels of atoms, such as those caused by heating or various other disturbances.

All active (moving) photons are endowed with parameters such as frequency (f), velocity (v), *free-space* wavelength (λ), and mass, and thus represent a finite unit of energy (e). In pure physics, the photon's energy, in joules per second, is usually defined by the formula,

$$e = h \times f$$

in which h is Planck's constant (= 6.626×10^{-34} J s).

In optoelectronics, it is more useful to define the energy in terms of electron-volt (eV) units, and to relate it to the photon's wavelength (l) in nanometres (nm) rather than its frequency. In this case the basic formula transforms into the easily remembered form,

$$eV = 1240/\lambda$$

Thus, a LED that generates a red output at a wavelength of 645nm has a bandgap energy value of 1.92 eV. The energy value of an individual photon

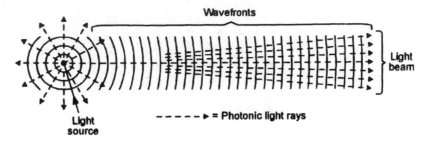

Figure 9.1   *Conceptual diagram illustrating some basic features of radiated light*

depends on its actual wavelength, but is very small; an ordinary green LED, for example, generates an output flux flow of about 2500 million photons per microsecond at a mean light output power level of a mere 1mW.

Figure 9.1 shows a simple conceptual diagram that illustrates some basic features of light when radiated from a small point source. The light flux (which contains vast numbers of photons) is effectively radiated in the form of a continuous series of spherical photonic waves that become progressively more planar (less sharply curved) as they move further from the source. The photons move outwards, perpendicular to the wave fronts; a photonic light *ray* traces the mean path of a photon; a photonic light-*beam* depicts the paths of a collection of rays. A light-beam is angular when close to the light source, but becomes progressively more parallel as the distance from the source increases.

In optoelectronics, the term 'light' relates to the entire visible light (400nm to 700nm) part of the electromagnetic spectrum, plus most of its invisible infra-red (IR) and ultra-violet (UV) ranges, i.e. to the spectrum's 10nm to 100mm section (see *Figures 1.1* and *1.2* in Chapter 1).

The Sun is the most powerful light generator in our solar system. It generates and radiates light energy as a byproduct of its continuously active nuclear fusion process; 60% of its radiant energy lays in the IR range. Only 0.0005% of the Sun's radiated energy is (after travelling a mean distance of 93 million miles through space in 8 minutes 20 seconds) received by planet Earth, and one-third of this reflects directly back into space. The energy contained in the remaining flux delivers a mean power of 4kW per square metre per day to the Earth's surface and acts as the engine for our planet's weather systems and (as a consequence of the results of photosynthesis, etc.) sustains all life on our planet.

Light travels through empty space at a velocity of 186 282 miles (299 792 kilometres) per second. Light's velocity was first *estimated* with reasonable precision by the Danish astronomer Ole Roemer in about 1690, after he observed unexpected variations in the actual and predicted times of the eclipse of Saturn's moons. The velocity of light through the Earth's atmosphere (which is 0.03 percent slower than through empty space) was first *measured* with reasonable precision (within 5 percent) by the French amateur scientist Armand Fizeau in Paris in 1849. Fizeau used an opto-mechanical stroboscopic technique to measure the time (about 57μs) that a beam of light took to cover a two-way 17km journey, and from the results estimated the speed of light at 313 300 kilometres per second.

The visibility of light

A stream of light (photons) racing through empty space can be regarded as a stream of latent energy, and is quite invisible; it only becomes visible when its flux strikes an absorptive material and releases some or all of its latent energy. These effects can be observed by looking up at the moon on a clear night; parts of its surface are illuminated because they are absorbing energy from the rays of the Sun, which is out of sight below Earth's horizon; the areas of space through which the sunlight is travelling appear completely dark.

If you go out into the open air on a bright summer's day, you will be bombarded by a stream of solar-generated IR, UV and visible light rays that will produce three distinct types of physiological effects on you. The IR rays will produce a sense of warmth wherever they strike exposed areas of your skin, and the UV rays will slowly start to change your skin's pigmentation in those exposed areas, eventually giving them a deep tan. The visible light rays from the Sun span the full colour spectrum. When they strike an object that you can see, the object's visual colouring is dictated by the object's spectral characteristics.

If the object that is exposed to the Sun's light absorbs all of the spectrum's light energy, the object appears black. If it absorbs only part of the available energy and reflects the rest, it will appear white if it reflects light equally across the entire spectrum, or red if reflects mainly the red part of the spectrum, or green if it reflects mainly the green part of the spectrum, and so on. The apparent colours have degrees of purity that depend on the width of the reflected part of the spectrum.

Note that human eyes do not have a linear spectral response (just as our ears do not have a linear aural response), and the response varies between individuals. The graph of *Figure 9.2* shows the spectral response of typical human eyes, which are ten times more sensitive to yellow-green (560nm wavelength) than they are to mid-blue (470nm) or mid-red (660nm). You can observe these effects by looking at a stained glass window (from within a building) when the window is brightly illuminated by the Sun; the window's glass segments all have roughly similar values of translucence, but the green segments seem far brighter than the mid-red or blue ones. The next section of this chapter gives more details on this subject.

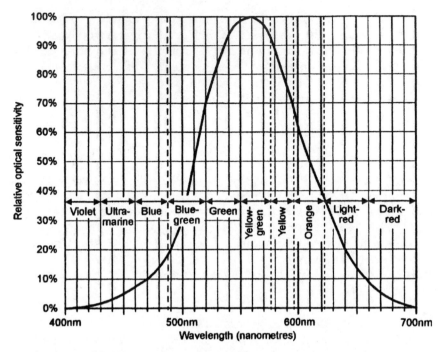

Figure 9.2   *Standardized relative spectral response of the human eye*

Light units

When dealing with light and optoelectronic components such as LEDs and lasers, etc., the units most often used in data sheets are those relating to the light's wavelength and spectral bandwidth, and to the intensity and power levels of its flux. Light *wavelength* is a measure of the light's colour; visible-light wavelengths fall within the range 400nm to 700nm; UV-light has a wavelength below 400nm; IR-light has a wavelength above 700nm.

The colour purity of a light is defined by its *spectral bandwidth*, which is measured between the points where the radiated power falls to half of its peak value. True white light contains all the colours of the 400nm to 700nm spectrum; it thus has a bandwidth of at least 300nm and is known as *chromatic* (multi-toned) light. Red LEDs (operating at about 650nm) have typical spectral bandwidths in the range 15nm to 50nm and are thus also chromatic, since their light outputs span various shades of red. Laser-generated light usually has an exceptionally narrow bandwidth (often less than 0.01nm), and is known as *mono-chromatic* (one-toned or pure-toned) or (if all of its emitted photons are in phase) *coherent* light.

Dealing next with the light units concerned with values of light intensity and power, it is important to note that – since human sight does not have a linear spectral response – these values may be expressed in two different types of unit. Conventionally, *photometric* units are used if they relate to the physiological (apparent) values of visible light sensed by humans, and *radiometric* units are used if they relate to genuine (true) values of visible or IR or UV light. The following four basic types of unit (each of which has both photometric and radiometric notations) are widely used in optoelectronics.

Total radiated flux power

Light is radiated energy; the total power of the flux flowing from a light source is measured in watts in radiometric notation, or in lumens in photometric notation. The photometric quantities are related to the corresponding radiometric ones by the internationally recognized standard 'human eyeball response' photopic conversion graph shown in *Figure 9.3*, which shows that 1 watt of light power is equal to 680 photometric lumens at a wavelength of 555nm (yellow-green), or roughly 82 lumens at 475nm (mid-blue) or 65 lumens at 660nm (mid-red), and so on.

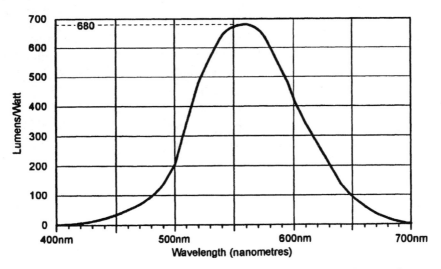

Figure 9.3 *Standard 'human eyeball response' photopic conversion graph showing the photometric lumens to radiometric watts relationship*

Flux density

In most practical optoelectronic applications, only a small fraction of a light source's total radiated power falls on a targeted light receptor such as a photocell or an eye's retina, and in such cases the most relevant parameter is the light's flux density (brightness) at the actual target point. In radiometric notation, this parameter is known as the light's *irradiance* value and is measured in watts per square metre (W/m²). In photometric notation the parameter is known as *illuminance* and is measured in lumens per square metre (lm/m²) or 'lux'. The lumen/watt relationship is the same as that shown *Figure 9.3*.

Angular flux intensity

Man-made light generators such as LEDs and filament lamps act like crude 'point' light sources but produce directional outputs, i.e. most of their available flux is concentrated into a cone of radiation. To specify flux intensity in such cases, a standard three-dimensional angular unit known as a steradian (symbol sr) is used; in radiometric measurements, angular flux intensity is known as *radiant intensity* and is specified in units of watts per steradian (W/sr); in photometric measurements, angular flux intensity is known as *luminous intensity* and is specified in units of *candela*, in which one candela equals one lumen per steradian (lm/sr).

Figure 9.4 shows a conceptual diagram that illustrates the basic features of a steradian unit. Imagine here that a point source of light is set at the centre of a translucent globe. From the point source, form a 57° cone that reaches out to the surface of the globe. This cone is a three-dimensional angular unit known as a steradian; the surface area of its mouth encompasses approximately 8% of the globe's total surface area.

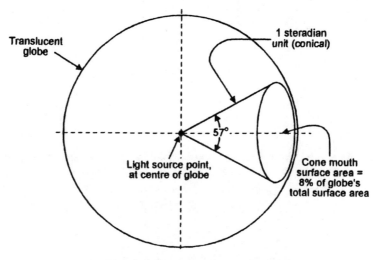

Figure 9.4   *Conceptual diagram illustrating the basic features of a steradian unit*

Radiated flux brightness

The brightness of a light source is proportional to both the radiated flux density and the radiating surface area of the light source. In radiometric notation, this parameter is known as the light source's *radiance* value and is measured in watts per steradian per square metre (W/sr × m²) of radiating surface area. In photometric notation the parameter is known as *luminance* and is measured in lumens per steradian per square metre (lm/sr × m²) or candela per square metre (cd/m²).

Light-beam manipulators

Visible and IR light beams can readily be reflected, bent, or manipulated in various other geometric ways with the aid of simple optical devices such as mirrors, retroreflectors, prisms or lenses. This section describes the basic operating principles and optoelectronic applications of such devices.

Mirrors

The simplest mirror is the ordinary flat totally reflective silvered-back glass type. If a narrow beam of light is aimed through the glass and onto the reflective (rear) surface of such a mirror, the reflected beam always obeys the basic *law of reflection*, which is illustrated in *Figure 9.5* and states that the angle of incidence (the angle between the arriving ray and an imaginary line drawn perpendicular to the mirror's surface) and the angle of reflection (the angle

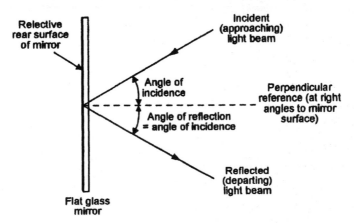

Figure 9.5 *Diagram illustrating the basic* law of reflection *that applies to a flat mirror*

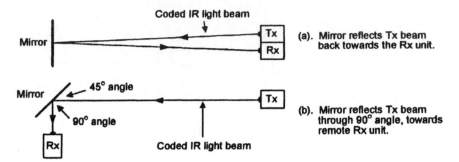

Figure 9.6   *Two simple ways of using mirrors in optoelectronic security applications*

between the reflected ray and the imaginary perpendicular line) are exactly equal.

Figure 9.6 shows two simple ways of using a mirror in optoelectronic security applications. In *Figure 9.6(a)* the mirror is used in a corridor protection system, to link a coded Tx IR light beam into an adjacent Rx unit, which activates an alarm if the beam is interrupted. In *Figure 9.6(b)*, the mirror is angled at 45° and projects the Tx beam around a 90° corner and on to a remotely placed Rx unit; this system can be used to protect an L-shaped corridor or two adjoining outside walls of a building, and is aligned by aiming the Tx beam directly at the mirror's Rx image.

Note that the image reflected by a simple mirror is reversed, left-to-right, as shown in *Figure 9.7*. If you take *Figure 9.7* and hold it in front of a mirror, you will see that all of its text is reversed in the reflected image, which is thus known as a 'virtual' (rather than real) image. While you are standing in front of the mirror, scratch your right ear; you will see that your virtual image is scratching its left ear.

If a mirror's reflection is viewed in a second mirror, the image that appears in the second mirror becomes real, rather than virtual. Try standing sideways

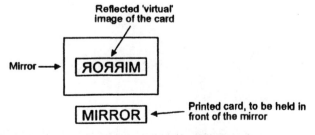

Figure 9.7   *The image reflected by a simple mirror is reversed, left-to-right*

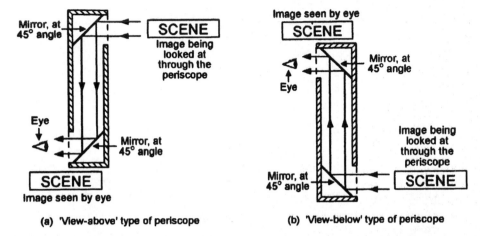

Figure 9.8 *A periscope presents a genuine (rather than 'virtual') image of a viewed scene*

in front of a large mirror, with a small mirror in your left hand; use the small hand mirror to view your image in the large mirror, and scratch your right ear; note that your image also scratches its right ear, but that if you look at your virtual images directly in either mirror it is the left ear that is being scratched.

One of the most common applications of the above 2–mirror technique is in simple periscopes, as illustrated in *Figure 9.8*. *Figure 9.8(a)* shows the basic construction of a conventional 'view-above' type of periscope. Here, the light from the scene that is being viewed strikes the upper mirror, is reflected downwards at an angle of 90° and strikes the face of the lower mirror, which bends the light through another 90°, where it can be viewed by the eye of the observer; the resulting image is real, rather than virtual, and is seen from a perspective above the viewer's eye level.

The 'view-above' type of periscope shown in *Figure 9.8(a)* can be made to function as a 'view-below' type of periscope by simply using it upside-down, as shown in *Figure 9.8(b)*. This periscope shows images from a perspective below the viewer's eye level; it is often used in the movie and TV industries to obtain ground-level shots of small animals or of miniature (model) towns or battle scenes, etc., for use in various films/videos.

Retroreflectors

A retroreflector is a passive device that automatically reflects a light's radiation back towards its source, irrespective of the light's precise angle of incidence. Devices of this type are widely used in reflective optoelectronic

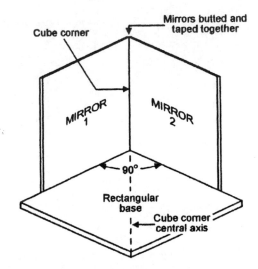

Figure 9.9 *Basic construction of a simple 'cube corner' retroreflector*

light-beam security alarms and barrier control systems and do not have to be precisely aligned with the light-beam source.

Figure 9.9 shows a way of using two small mirrors (or mirror tiles) to make a device known as a cube corner retroreflector. The two mirrors are simply butted and taped together and set at an angle of 90° by pressing them against the edges of a rectangular base (such as a soft-cover book). If a light beam (or image) is aimed into the cube corner of this device from any point that is at right angles to the cube's vertical plane and anywhere within ±35° of its central corner axis on the horizontal plane, the device automatically reflect the light (or image) back towards its source point. *Figure 9.10* illustrates the unit's operating principle.

In *Figure 9.10(a)*, the light-beam is projected from a left-of-centre source towards the retroreflector's corner, hits the LH mirror at an incident angle of (say) 75°, is reflected off again at 75°, hits the RH mirror at an incident angle of 15°, and is reflected off again at an angle of 15°, and heads back towards the source on a parallel path. The light-beam's total angular change is equal to the sum of the two incident and two reflection angles, and inevitably equals 180°. This same basic action is obtained in *Figures 9.10(b)* and *(c)*, except that different incident and reflection angles are involved and that the *Figure 9.10(c)* illustration shows the beam bouncing from the RH mirror on to the LH one.

Note in *Figure 9.10* that an imaginary line drawn centrally between the transmitted and returning parallel beams always hits the cube corner, and that the two beams thus lay symmetrically about this line. This action enables the cube corner retroreflector to produce some unusual visual effects when viewed in frontal elevation, as illustrated in *Figure 9.11*.

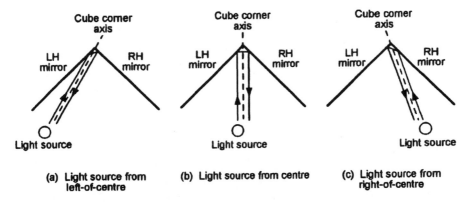

Figure 9.10 *Diagram illustrating the functional operation of the cube corner retroreflector*

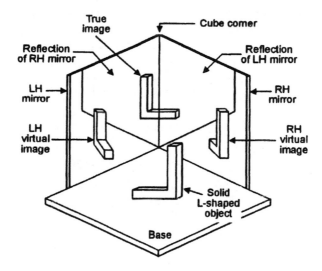

Figure 9.11 *When viewed in frontal elevation, this retroreflector produces three images of an object*

In *Figure 9.11*, a three-dimensional solid L-shaped model is turned around so that it is facing the retroreflector and is placed in front of its cube corner. Note that the LH mirror reflects the RH mirror, and vice versa, producing a 'mirror cube' reflection. The retroreflector produces virtual images of the object in both the LH and RH mirrors, and produces a true image of the object in the mirror cube.

The cube corner unit gives only one-dimensional (horizontal plane) retroreflection. An alternative design is the trihedral unit, which gives two-

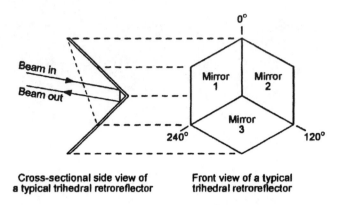

Cross-sectional side view of a typical trihedral retroreflector

Front view of a typical trihedral retroreflector

Figure 9.12   *Basic construction of a trihedral retroreflector*

dimensional (vertical and horizontal planes) retroreflection of light beams. *Figure 9.12* shows the basic construction of this unit, which uses three diamond-shaped mirrors set at 120° to each other; the unit's action is such that a light-beam entering its front is reflected through 180° in two dimensions by the mirror surfaces and then returns towards the source point on a parallel path. Often, hundreds of miniature retroreflectors of this basic type are used in road-side signs, making them glow brilliantly in the headlights of passing vehicles.

Prisms

In optics, a prism is a block of transparent material having two or more plane (flat) surfaces. Prisms have an innate ability to bend the paths of light. A flat sheet of glass is a very simple prism.

When light travels through a transparent material or medium other than free space, its velocity ($v1$) and wavelength ($\lambda1$) are lower than those that pertain in free space ($v0$ and $\lambda0$). The ratio $v0/v1$ or $\lambda0/\lambda1$ is known as the medium's *refractive index*, and is notated by the symbol *n*. *Figure 9.13* lists

Substance	Refractive Index
Free Space	1.000 000
Air	1.000 293
Ice	1.31
Water at 20°C	1.333
Glass	1.5 to 1.9
Diamond	2.42

Figure 9.13   *Table listing typical refractive index values for various transparent media*

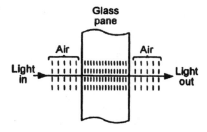 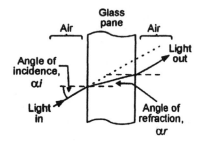

(a) A light ray and its wavefronts passing through a sheet of glass after arriving at an incident angle of zero degrees.

(b) A light ray passing through a sheet of glass after arriving at an incident angle of about thirty degrees.

Figure 9.14   *Diagrams illustrating the effects of refraction through a sheet (pane) of glass*

some typical n values for various transparent media, including normal glass ($n = 1.5$ to 1.9), which is transparent to all visible light and much of the lower IR spectrum, but blocks most of the UV spectrum.

Figure 9.14 illustrates the effects that 'refraction' has on a light ray when it travels through a pane of glass with (for simplicity) an n value of 2. In *Figure 9.14(a)* the beam enters the glass at an incident angle of $0°$, is slowed down (thus reducing the light's wavelength but *not* its frequency) as it passes through the glass, and then returns to normal 'though air' speed as it leaves the glass at an angle of $0°$.

In *Figure 9.14(b)* the dashed lines represent the *normal* or zero degrees angular reference line. In this diagram the ray enters the glass at an incident angle (αi) of (say) $30°$ and then, as it passes from the thinner medium (air) to the denser one (glass), the ray *bends towards the normal* by a refractive angle (αr) of about $15°$; when the ray passes from the denser to the thinner medium again as it leaves the glass, the ray *bends away from the normal* again, returning to its original angle of $30°$, thus obeying the basic Laws of Refraction. The values of αi, αr, and the n values of the incident (ni) and refractive (nr) regions are related by Snell's law of refraction, which states that $nr/ni = \sin \alpha i / \sin \alpha r$.

Note in *Figure 9.14(b)* that the light ray leaving the glass is shown parallel with, but offset from, the path of the original input ray, which is indicated by the dotted line. The degree of offset (parallax error) increases with the angle of incidence and with the thickness of the glass. Simple mirrors, in which light rays enter and leave the mirror via the same glass surface, are subject to this type of parallax error.

Most prisms have plane surfaces that are angled away from each other, as shown in *Figure 9.15*, in which two prisms each have their major surfaces angled at $30°$ to one another. This diagram shows the effects that the prisms have on a light ray that arrives at an incident angle of zero degrees. In both

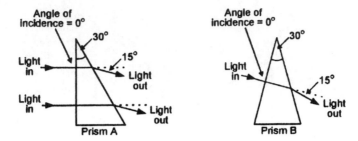

Figure 9.15 *Diagram illustrating some important ray-bending features of a simple prism*

cases the ray passes cleanly through the glass without bending, but on leaving the glass the ray bends away from the normal by about 15°, thus obeying the basic Laws of Refraction. Note that the degree of ray bending is independent of the thickness of the prism glass and (ignoring n value effects) is determined mainly by the angle of incidence of the ray and by the angular difference between the prism's input and output surfaces.

Regarding the 'n value effects' mentioned in the above paragraph, it is important to note that the refractive index values of all transparent media other than free space vary with the wavelength of a light, and increases as wavelength shortens. The refractive index of glass is normally measured using a yellow sodium light with a wavelength of 589nm; the actual index value is higher than normal to violet light (400nm wavelength) and lower to red light (700nm wavelength).

Figure 9.16 shows, in exaggerated form, the results of passing a narrow beam of white light through a symmetrical triangular prism. White light contains all the colours (wavelengths) of the visible spectrum, and the prism thus (because its refractive index is wavelength-dependent) bends each individual colour of the beam by a different amount, giving the least bend to red light and the greatest bend to violet light. The prism's output thus

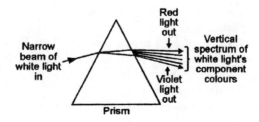

Figure 9.16 *Exaggerated diagram showing a prism splitting a beam of white light into its component colours*

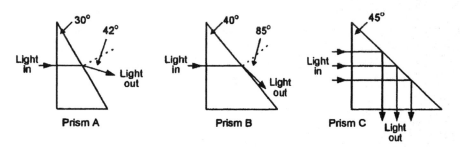

Figure 9.17 *Diagram illustrating the phenomenon of total internal reflection in a prism*

takes the form of a vertically-expanded coloured spectrum. This scattering of white light's component colours is known as *dispersion*.

When a ray of light passes through air and enters a prism, it bends by an amount determined by its angle of incidence and by the refractive index value of the glass. When the ray leaves the prism again and returns to the air, it bends by an amount determined by its angle of incidence and by the refractive index of the air (1.0) divided by that of the glass (say 1.5), and this value is invariably less than zero (0.667 in this example). *Figure 9.17* shows the actual amounts of output refraction that occur on three different prisms that each have a refractive index of 1.5.

In *Figure 9.17*, the ray strikes the output surface of prism A at an incident angle of 30° and leaves the prism at a refractive angle of 42°; this prism thus bends the ray downwards by 12°. In the case of prism B, the ray strikes its output surface at an incident angle of 40° and leaves at a refractive angle of 85°, thus bending the ray downwards by 45°.

Note in the case of prism B that the ray leaves the prism at an angle that is only 5° less than the angle of slope of the prism's output face, and it is obvious that if the angle of incidence is increased much more the ray will be unable to penetrate the prism's output surface. The angle of incidence at which this occurs is known as the surface's *critical angle*, and is dictated by the *n* value of the glass; the critical angle is 43° at an *n* value of 1.5, 36° at an *n* value of 1.7, and 32° at an *n* value of 1.9.

Figure 9.17 shows, in the Prism C diagram, what happens to the light rays when they strike the prism's output face at an incident angle of 45°, i.e. at an angle greater than the critical angle of the surface. Under this condition a phenomenon known as *total internal reflection* occurs and makes the internal surface act like a mirror that bends the rays by double their angle of incidence, thus (in this case) bending them through a 90° angle and projecting them through the lower face of the prism.

The internally reflecting type-C prism thus acts like a mirror that bends light through 90°, but (since the light passes through separate input and output surfaces) does not suffer from parallex errors. This type of prism is widely used in high quality optical instruments and devices such as reflex cameras, binoculars, and automatic laser-aiming controllers.

Total internal reflection can occur whenever one tranparent material interfaces with another that has a lower refractive index, thus giving a less-than-unity refractive index and a positive critical angle value at the interface junction. All modern fibre optic cables rely on this 'total internal reflection' principle for their very efficient low-loss operation (fibre optic cable principles are described in Chapter 10 of this volume).

Lenses

Normal optical lenses are light-bending refractive devices that are related to prisms but have curved (rather than flat) faces. *Figure 9.18* shows the classic profile of a simple lens that has two parallel faces that are each radially curved in two dimensions, to form a section of a sphere. This type of lens can focus a parallel bundle of light rays onto a single point (the *focal point*), as shown in *Figure 9.18(a)*; the distance between the centre of the lens and the focal point is the *focal length* of the lens. If a light point-source is spaced from the lens by a distance equal to the focal length of this type of lens, the lens converts the light into a parallel ('collimated') beam of light, as shown in *Figure 9.18(b)*.

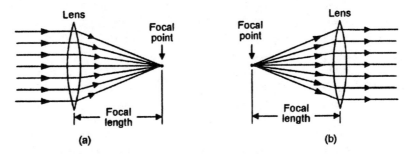

Figure 9.18 *A simple lens can (a) focus a parallel bundle of rays onto a single point, or (b) convert the rays from a single point source into a parallel (collimated) beam of light*

Figure 9.19 shows the basic way of using two simple lenses in a light-beam alarm or communication system. At one end of the system, the left-hand lens converts the tranmitter's light point-source into a collimated (parallel) light beam, and at the other end of the system the right-hand lens converts the collimated beam back into a point of light, which is applied to the receiver's

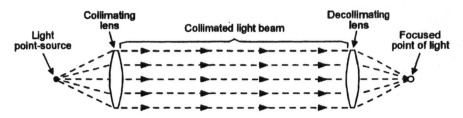

Figure 9.19 *Basic way of using two lenses in a light-beam system*

light-sensitive input. Identical lenses are used in the collimating and decollimating processes.

Systems of the above type generate a perfectly parallel beam only if the light point-source is infinitely small, and this is an impossibility. In practice, the beam widens with distance after leaving the collimating lens; the amount of widening is proportional to the width of the light source and inversely proportional to the diameter of the lens. For minimum widening, the lens diameter must be large relative to that of the source.

Simple lenses with one or both faces shaped as a section of a sphere are known as *spherical* lenses and are available in *convergent* and *divergent* types; convergent types make a parallel beam of light converge towards a common focal point; divergent types make a parallel beam of light diverge outwards. *Figures 9.20* and *9.21* show a variety of lenses of these types.

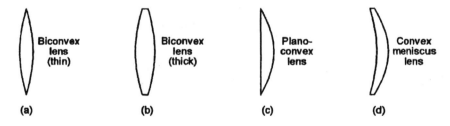

Figure 9.20 *Four simple types of convergent spherical lens*

Figure 9.20 shows four simple types of convergent spherical lens. The thin biconvex lens shown in *(a)* is the same type as shown in *Figure 9.18*; it operates equally well either way around, but its sharp edges are rather fragile. The thick biconvex type shown in *(b)* gives the same performance as the *(a)* type, but is more rugged. The plano-convex type shown in *(c)* has one flat and one curved face and must be used the correct way around, with the flat face pointing in the direction of the parallel light beam and the curved face aimed towards the light's focal point. The convex meniscus type shown in *(d)* has a very long focal length; it is the type used in most spectacles and contact lenses.

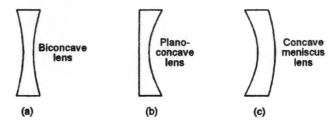

Figure 9.21   *Three simple types of divergent spherical lens*

Figure 9.21 shows three simple types of divergent spherical lens. The biconcave lens shown in *(a)* can be used either way around, but the plano-concave type shown in *(b)* must be used the correct way around. A concave meniscus type is shown in *(c)*. Lenses of these various types are often used in conjunction with convergent lenses, to make high-quality compound lenses (as described shortly).

Most spherical lenses are of the simple type already shown, but other types are also available. The cylindrical biconvex lens shown in *Figure 9.22(a)*, for example, is curved in one dimension only, and is used to focus a parallel light beam into a thin line, rather than a spot, or to convert a thin line of light into a collimated beam.

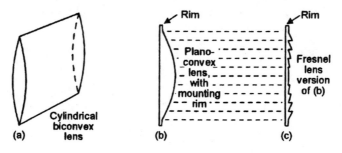

Figure 9.22   *Three special types of spherical lens*

In most normal lenses much of the sub-surface glass or plastic lens material performs no useful function. A *Fresnel* lens is one in which most of the non-useful lens material is removed, to make a lightweight or low-cost lens that performs almost as well as a normal lens in simple 'light-beam' types of application. *Figures 9.22(b)* and *(c)* show, in cross-section form, how a plano-convex lens with a mounting rim is transformed into its Fresnel equivalent. Here, the Fresnel lens is made up by effectively dividing the original *(b)* lens into ten horizontal slices, removing the 'dead' material from each slice, and then bonding the remainder onto a base of identical material that also acts

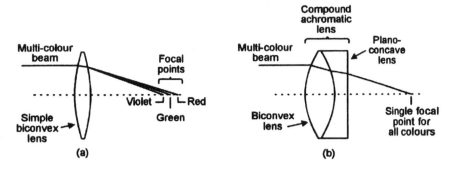

Figure 9.23 *The compound achromatic lens shown in (b) is designed to minimize chromatic aberration of the type suffered by the simple (a) lens*

as the lens rim. In practice, most Fresnel lenses are moulded, in plastic or glass.

The focused images generated by simple spherical lenses suffer from spherical and chromatic defects or abberations. *Spherical abberation* makes a straight line appear curved in the focused image; if you wear spectacles, you can see a demonstration of this effect by standing in front of a set of library shelves and noting how the shelves above and below your eye level appear curved when you have your glasses on, but not when they are off.

In simple lenses, *chromatic abberation* makes faint coloured fringes appear around focused white or multi-coloured images. The effect occurs because the refractive index of the lens material (and thus the focal length of the lens) varies with the colour of light, as shown in *Figure 9.23(a)*, making it possible to sharply focus only a small slice of the colour spectrum; the rest of the spectrum is out of focus, producing the 'fringe' effect. This problem can be overcome by using a compound lens made of one converging and one diverging lens, each with a different refractive index value, as shown in *Figure 9.23(b)*. Such a lens can be made to give all colours the same overall focal length, and is known as an *achromatic* or *antispectroscopic* lens.

The strangest and most recently developed lens is the *graded-index* (GRIN) *rod lens*, which is used in modern fibre optic applications and operates in a different way to a normal lens. A GRIN *rod* is a glass or fibre rod that has a refractive index that decreases progressively with distance from the rod axis. This index variation causes light rays to follow a sinusoidal path as they travel along the rod, as shown in *Figure 9.24(a)*. The length of one complete sinusoidal cycle in the rod is called the *pitch* (P) of the rod; the P value is determined mainly by the rod's diameter and refractive index profile; the value is typically about 20mm.

A GRIN rod *lens* is simply a slice of GRIN rod with a length less than a single pitch-length, so that its optical output signal is out of phase with the optical input signal. The most interesting and widely used GRIN rod lens has

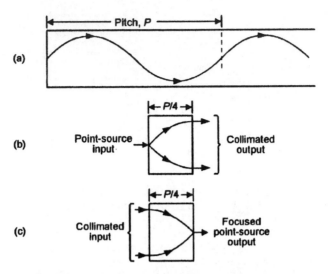

Figure 9.24   *Light rays follow a sinusoidal path through a graded-index (GRIN) rod (a). A quarter-pitch GRIN lens collimates the light from a point-source (b), and focuses a collimated light beam into a point (c)*

a quarter pitch (giving it a length of about 5mm) and has the interesting properties illustrated in *Figures 9.24(b)* and *(c)*. Note in *(b)* that the light from a point source in contact with the centre of the left face of the lens emerges from the right face in fully collimated form, and in *(c)* that a colli-mated light beam entering the left face leaves the right face as a focused spot. The GRIN rod lens thus has some properties of a conventional lens, but has a very short focal length.

   GRIN rod lenses are widely used in modern fibre optic and laser module applications. *Figure 9.25* shows an example that illustrates the advantages of a GRIN rod lens over a spherical lens in a fibre optic application in which the fibre optic cable's point-source 'light' output needs to be collimated when

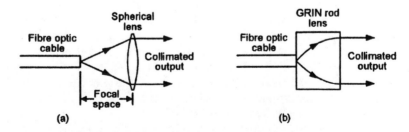

Figure 9.25   *Method of collimating the light output from a fibre optic cable using (a) a spherical lens and (b) a GRIN rod lens*

fed into the outside world. In the case shown in *(a)*, the light is collimated by a conventional lens, which must be placed a precise fixed distance from the end of the cable, to which it is coupled by an air gap. In the case shown in *(b)*, the GRIN rod lens is simply bonded directly to the polished end of the cable, and no carefully spaced air gap is required. More information on fibre optic and GRIN (graded index) operating principles is given in Chapter 10.

Fibre optics and lasers

The two most revolutionary and important optoelectronic-associated events to have occurred since the mid-1960s are the development of practical fibre optic cables for use in interference-free wide-band communication systems, and the development of reliable long-life low-cost laser units for use in every-day gadgets such as CD-players and bar-code readers. The first half of this chapter explains the basic operating principles and applications of fibre optics; the second half gives similar treatment to LEDs and lasers.

Fibre optics

Introduction

A fibre optic cable can, in very simple terms, be regarded as a hair-thin flex-ible light pipe or *optical waveguide* that can efficiently carry modulated or unmodulated light signals from one point to another (even if the journey involves bends and loops) with complete immunity from electromagnetic interference. In most practical applications the cable must be linked to the light source and destination points via special connectors, to minimize signal losses.

Figure 10.1 shows the basic elements of a simple one-channel one-way fibre optic communication system, in which the output of the optogenerator LED is pulse-modulated and is coupled to the input of optosensor Q1 and its pulse-detecting unit via a length of sheathed fibre optic cable.

The *Figure 10.1* circuit can send data from point A to point B, but not the other way around. *Figure 10.2* shows a 2–way (bidirectional) version of the above circuit. Here, signals can be sent from A to B via Tx unit 1, cable 1, and Rx unit 1, and from point B to point A via Tx unit 2, cable 2, and Rx unit 2. In practice, special 2–core 'duplex' fibre optic cables are available for use in this type of application.

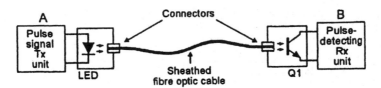

Figure 10.1 *Simple single-channel 1–way fibre optic communication system*

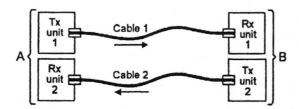

Figure 10.2 *Simple single-channel 2–way fibre optic communication system*

The *Figure 10.1* and *10.2* circuits are designed for use in applications in which each fibre optic cable carries only a single data channel. If multi-channel operation is needed, the channels can be electronically multiplexed, or each channel can be superimposed in a high-frequency carrier wave in the manner of a normal radio or TV channel.

When the commercial possibilities of fibre optic cables were first postulated in the early 1960s, there was much speculation regarding the probable performance capabilities of such cables. Visible light has, for example, a frequency spectrum ranging from 430GHz (700nm wavelength) to 750GHz (400nm wavelength), and it was presumed that fibre optic cables would have bandwidths that would easily cover this 320GHz spectrum, thus enabling them to handle up to 64 000 5MHz data channels (Note: 1GHz = 1000MHz). In practice, modern fibre optic cables fall well short of this imagined ideal; many have real-life bandwidths of only a few hundred megahertz, and can handle only a few dozen 5MHz data channels.

In spite of their limitations, fibre optic communication systems outperform conventional wire-connected systems in many important respects. They are immune to electromagnetic interference, are electrically fully isolated, are low in weight, offer high security, and (unlike copper wires) are made from a cheap and very abundant raw material (silica).

Fibre optic principles

An optical fibre is a wire-like strand of high-purity glass or plastic in which the refractive index of the outer layer is lower than that of the inner core,

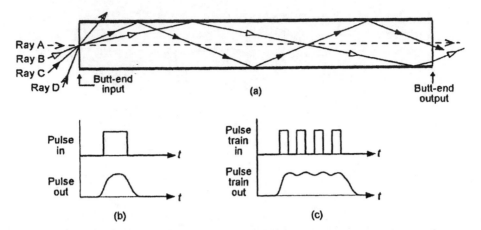

Figure 10.3 *Diagram illustrating basic 'step-index' fibre optic operating principles*

thus endowing the fibre with the optical phenomenon of total internal reflection (as described in Chapter 9). *Figure 10.3* illustrates some elementary fibre optic operating principles and helps explain some basic fibre optic jargon. The main (a) part of the diagram shows a simple glass rod or 'light pipe' in which the glass has a refractive index of 1.5; light rays travel through this glass at a speed of 200 metres/μs.

Imagine in *Figure 10.3* that a series of light-beam pulses are fed into the butt-end input of the glass rod via a focusing lens. The beam is made up of a vast number of photonic rays, which enter the butt-end input at a variety of angles of incidence. The diagram depicts the paths of four such rays, which travel in the following individual ways.

Ray A strikes the butt-end input of the glass rod at an angle of incidence of zero degrees, passes cleanly through the glass rod via the shortest possible route, and emerges from the rod's butt-end output after travelling through the rod at a mean speed of 200 metres/μs.

Ray B strikes the butt-end input at a moderately sharp angle of incidence, is refracted to an angle of 12° on entering the glass, and after travelling a fairly short distance strikes the rod's upper glass-to-air boundary, where (because it arrives at a fairly shallow angle) it is subjected to the phenomena of total internal reflection (fully described in Chapter 9), which reflects the ray downwards at an identical angle until it strikes the rod's lower glass-to-air boundary, from which it is reflected upwards again, and so on as the ray *propagates* its way along the rod until it eventually leaves its butt-end output. The ray thus follows the repeatedly reflected zig-zag course shown in the diagram; this beam travels a greater distance than ray A in its journey through the glass rod, actually travelling through it at a mean horizontal speed of 195 metres/μs.

Ray C, on entering the rod, is refracted to an angle of 22° on entering the glass and then propagates its way along the tube in the same basic way as ray B, but does so at sharper reflective angles and thus takes even longer than ray B to complete its journey through the glass rod, actually travelling through it at a mean speed of 185 metres/μs.

Ray D strikes the butt-end input at a very sharp angle of incidence, is refracted to an angle of about 60° on entering the glass, and after travelling a short distance strikes the rod's upper glass-to-air boundary, where (because it arrives at an angle greater than the critical angle of the boundary) it simple passes through the boundary and is dissipated into the air. This ray thus fails to propagate along the glass rod.

Note that *Figure 10.3* depicts the fibre optic rod's operation in a simplified 2-dimensional style, in which the rays simple bounce off the internal top and bottom faces of the rod. In reality, rays travel 3-dimensionally, bouncing off the rod's top, side and bottom faces, and thus tend to spin or skew as they travel along the rod.

Fibre optic jargon

Various 'jargon' words and phrases are widely used in fibre optics and have meanings that can be explained with the aid of *Figure 10.3*, as follows.

Step index. Fibre optic cable operation depends on the total internal reflection phenomenon that can occur at the points where a transparent material meets a material or substance with a refractive index lower than its own. In the *Figure 10.3* example, the glass rod has a refractive index of 1.5, and the surrounding air has a refractive index of 1.0, and reflection can occur at the precise point where the ray meets the abrupt or *step* reduction of the refractive index. Cables of this type are called *step-index* fibre optic cables. *Graded-index* (GRIN) cables are also available, and are described later in this chapter.

Maximum acceptance angle and *numerical aperture*. Light rays will only propagate through a fibre optic cable if the input ray applied to the cable's butt-end input falls within a certain angular range. In *Figure 10.3*, for example, rays A (arrival angle = 0°), B (24°), and C (45°) all propagate, but ray D (75°) does not propagate. The maximum input angle at which propagation will occur is called the cable's *maximum acceptance angle*, and is denoted by the symbol αM. This basic parameter is often expressed in sine form, in which case it is called the cable's *numerical aperture*, NA, and equals sin αM. Thus, a cable with a maximum acceptance angle of 32° has a numerical aperture of 0.53.

Operating mode. In *Figure 10.3*, light rays can propagate along the glass rod at a variety of angles or 'operating modes'. Rays that propagate at steep reflective angles travel in *high-order mode*, and ones that propagate at shallow angles travel in *low-order mode*. Since the rays can travel in a variety of modes in this particular fibre optic rod, it is called a *multi-mode* or MM device. Some fibre optic cables can operate in only a single mode, and are called *single mode* (SM) or *monomode* devices.

Modal dispersion. When light rays propagate through a step-index multi-mode fibre optic cable or rod, their speed of linear travel depends on their angular operating mode. High-order mode rays travel at lower linear speeds than low-order ones. In the *Figure 10.3* diagram, for example, ray A travels at a linear speed of 200 metres/μs and takes 5ns to travel 1 metre; the slowest (highest-order) ray that can propagate through this particular rod travels at 141 metres/μs and takes 7.1ns to travel 1 metre.

Consequently, if a very fast sharp-edged light pulse is fed to the input of the *Figure 10.3* fibre optic rod, the individual photonic rays representing its leading or trailing edge at the rod's output are spread over a 2.1ns time spread in a 1 metre rod, or 21ns in a 10 metre rod, and so on. This 'spreading' is called *modal dispersion* and produces a degraded (blurred) output pulse, as shown in *Figure 10.3(b)*. Similarly, if a train of ultra-fast pulses are fed through the rod, modal dispersion may degrade them into the form of a single pulse, as shown in *Figure 10.3(c)*.

Modal dispersion is the major factor that limits the maximum pulse-operating rates (frequencies) of fibre optic cable, and is determined both by the design and the length of the cable. Cables with low modal dispersion figures offer higher maximum pulse operating rates than cables with high modal dispersion figures.

Material dispersion. The speed at which light travels through a fibre optic material depends upon the material's refractive index, and this varies with the light's wavelength. All real-life light sources have finite spectral bandwidths, and their different component wavelengths thus travel through the material at different speeds; such light thus undergoes spectral dispersion (spreading) as it travels through the fibre. This phenomenon is known as *material dispersion*, and, like modal dispersion, limits the maximum pulse-operating rates of fibre optic cable. Its degrading effects are directly proportional to the light source's spectral bandwidth, but are normally small compared to those caused by modal dispersion. Material dispersion can typically be reduced by a factor of about 200 by using a laser diode, rather than a normal LED, as the light source that feeds signals into the fibre optic cable.

Attenuation. When light passes through any translucent glass or plastic material its power is inevitably attenuated in the process. The amount of attenuation is directly proportional to the thickness (or length) of the material, and is measured in terms of dB/metre or dB/km. The high-quality glass used in spectacle lenses has a typical attenuation value of 6dB/metre (6000dB/km).

By contrast, the ultra-pure silica (silicon dioxide) glass used in glass fibre optic cables has a typical attenuation value of less than 0.0013db/metre (1.3dB/km) when used under optimum conditions.

Four basic mechanisms are primary causes of light attenuation in glass. Three of these (electron absorption, Rayleigh scattering, and material absorption) are *intrinsic absorption mechanisms* inherent in the basic nature of glass; the fourth mechanism (impurity absorption) is an *extrinsic absorption mechanism* caused by contamination (impurity) within the manufactured glass.

In ordinary glass, the extrinsic (impurity) absorption is the dominant mechanism in the basic attenuation process. In modern fibre optic cables, the impurities in the silica glass are reduced to the lowest commercially viable levels, and this results in the type of attenuation graph shown in *Figure 10.6*, which in turn is derived by combining the intrinsic and extrinsic absorption graphs shown in *Figures 10.4* and *10.5*. Note that these particular graphs are based on typical cable of the late 1980s to early 1990s era, and that most modern cables give attenuation figures that are somewhat lower than those shown.

The *Figure 10.4* graph shows the typical attenuation that occurs as a result of the intrinsic absorption mechanisms inherent in the basic nature of silica glass. Here, *electron absorption* occurs when photons are absorbed by individual atoms in the silica glass, raising one of the atom's electrons to a high-energy (excited) state in the process. Electron absorption causes heavy ultra-violet absorption that peaks at 140nm but extends well beyond 400nm wavelengths.

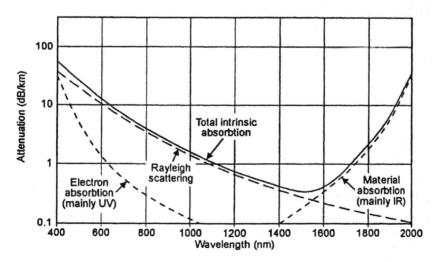

Figure 10.4 *Graph showing typical attenuation due to intrinsic optical absorption in high-grade silica fibre optic glass cables of the late 1980s to early 1990s era*

The *Rayleigh scattering* shown in *Figure 10.4* occurs when photons meet molecule-sized density irregularities in the glass and undergo shifts in their refraction angles. The consequent attenuation effects are severe when the photon's wavelength is similar to the dimension of the irregularity, but decline steadily as wavelength increases above 400nm.

The *material absorption* shown in *Figure 10.4* occurs when photon energy is absorbed by the atomic/molecular bonds associated with the basic glass meterial, and causes heavy attenuation of infra-red at wavelengths greater than 2000nm. The net result of the effects of the three intrinsic absorption mechanisms is that the optical attenuation of the glass is high (70dB/km) at 400nm, falls smoothly to a minimum value of about 0.4dB/km at 1500nm, and then rises rapidly again (because of material absorption) at longer wavelengths.

The *Figure 10.5* graph shows the basic form of the attenuation that occurs as a result of extrinsic absorption mechanisms in most high-grade normal-quality silica fibre optic glass. These losses are caused primarily by the presence of hydroxyl (OH) ions in the glass, which result in three large peaks of absorption within the 400nm to 2000nm spectrum. The largest peak occurs at 1370nm, and the other two occur at 950nm and 725nm; a very minor additional peak occurs at 1230nm.

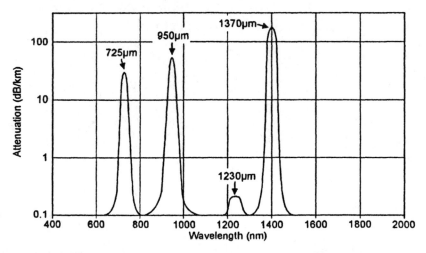

Figure 10.5 *Graph showing typical attenuation due to extrinsic optical absorption in high-grade silica fibre optic glass cables of the late 1980s to early 1990s era*

The *Figure 10.6* graph shows the typical attenuation that occurs as a result of all intrinsic and extrinsic absorption mechanisms in high-grade normal-quality silica fibre optic glass. Important points to note from this particular

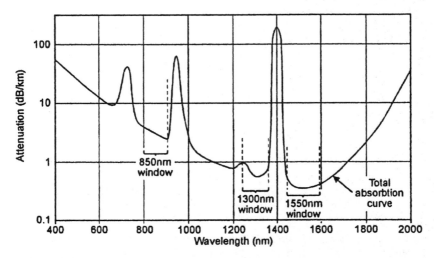

Figure 10.6 *Graph showing typical attenuation due to all forms of optical absorption in high-grade silica fibre optic glass cables of the late 1980s to early 1990s era*

graph are that attenuation is very high (greater than 10dB/km) over most of the visible-light (400nm to 700nm) spectrum and in various other areas, and that there are three 'window' areas in the infra-red range (at 850nm, 1300nm and 1550nm) at which attenuation figures attain local 'low' values. Practical glass fibre optic communication systems are nearly always operated within one or other of these 'window' ranges.

Polymer cable. The most widely used type of fibre optic cable is made from silica glass and can only be correctly cut with a special tool. This type of cable is meant for use with IR light, primarily in medium- to long-haul (greater than 250 metres) communication and data-transfer systems. There is, however, an alternative and cheaper type of fibre optic cable; it is made from plastic polymer and can easily be cleanly cut with a sharp blade, and is best suited to use with visible red (rather than IR) light, in very-short-haul (up to 50 metres maximum) applications.

Figure 10.7 shows the typical attenuation graph of a modern polymer fibre optic cable. This particular cable gives a mean attenuation of about 200dB/km (0.2dB/metre) when averaged over the full visible light spectrum, but about 1500dB/km (1.5dB/metre) when used in the low infra-red ranges.

Practical fibre optic cable types

The simple glass type of fibre optic rod or cable shown in *Figure 10.3* is too crude to be of real practical value, since its internally reflective glass-to-air

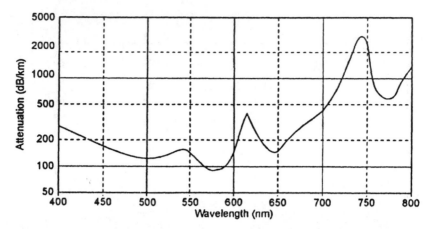

Figure 10.7   *Graph showing typical attenuation due to all forms of optical absorption in modern plastic polymer fibre optic cable*

junction is highly susceptible to external damage from scratches and contamination, and the very large differences between the refractive index values of its two media (about 1.5 for glass, 1.0 for air) give excessively large modal dispersion values.

All practical, modern, fibre optic cables are designed to give low values of modal dispersion, and use either a multi-mode or single-mode step-index form of construction, or a graded-index (GRIN) form of construction, as shown in *Figures 10.8* to *10.10*, which also show the basic refractive index (*n*) profiles and typical dimensions of their 'glass type' fibres. Each of these fibres has a solid core that is surrounded by a solid cladding with a refractive index value lower than that of the core; this cladding makes the cable's modal dispersion value immune to the effects of external surface scratches, etc. The outer core is further enclosed in an opaque primary outer sheath that provides further physical protection and immunity to optical interference.

Descriptions of the three basic cable types are given below; when reading these descriptions, remember that a human hair has a typical thickness of about 50μm, and note that fibre optic cable basic dimensions are conventionally notated in an 'a/b/c' or 'a/b' form, in which 'a' is the outside diameter of the fibre core in μm, 'b' is the outside diameter of the fibre cladding in μm, and 'c' is the outside diameter of the cable's primary outer sheath in μm. Thus, a 50/125/250μm cable uses a 50/125μm fibre that is protected by a 250μm-diameter primary sheath.

Multi-mode step-index. Figure 10.8 shows basic details of this type of fibre optic cable, in which a relatively thick core (typically 50μm or 62.5μm in glass fibres) is covered by a layer of cladding with a refractive index value that is

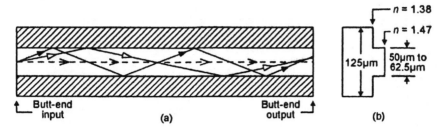

Figure 10.8 *Basic details (a), plus refractive index profile and typical dimensions (b), of the glass fibre used in a modern multi-mode step-index fibre optic cable*

only a few percent lower than that of the core, thus giving the fibre's butt-end input a fairly narrow (typically 11° to 24° maximum) acceptance angle and thus endowing the fibre with fairly low values of modal dispersion. These characteristics and dimensions make it easy to couple optical signals into and out of this type of cable. The most popular 'glass' versions of this cable have basic fibre dimensions of 50/125μm or 62.5/125μm. Typical bandwidth–length product values vary from 1MHz/km for plastic cables to 33MHz/km for glass types; these cables are thus used mainly in short- to medium-haul low-data-rate applications.

Single-mode step-index cables. Figure 10.9 shows basic details of a typical single-mode step-index fibre optic cable, in which light travels through the core only in the very shallow 'low-order' mode. Here, a very thin (5 to 8μm) glass core is covered by a fairly thick layer of cladding with a refractive index value only slightly lower than that of the core, thus giving the fibre's butt-end input a very narrow (less than 6° maximum) acceptance angle and endowing the fibre with a very low modal dispersion value. These characteristics and dimensions make it difficult to couple optical signals into and out of the cable, which require great precision in cutting and coupling. Never-

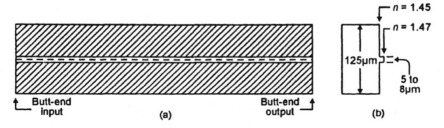

Figure 10.9 *Basic details (a), plus refractive index profile and typical dimensions (b), of the glass fibre used in a modern single-mode step-index fibre optic cable*

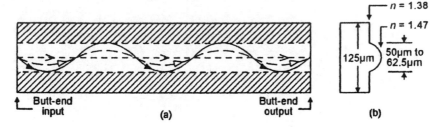

Figure 10.10   *Basic details (a), plus refractive index profile and typical dimensions (b), of the glass fibre used in a modern graded-index (GRIN) fibre optic cable*

the-less, the great bandwidth – length product values of this type of cable (typically between 1GHz/km and 10GHz/km) make it the most widely used type in long-haul high-data-rate applications.

Graded-index (GRIN). Figure 10.10 shows basic details of a 'GRIN' type of fibre optic cable. Here, the core has a refractive index value of about 1.47 along its central axis, but the value gradually diminishes (in 'graded' form) as the distance away from the central axis increases, until it falls to a minimum value of (say) 1.38. Photonic rays that arrive at the cable's butt-end input at fairly sharp angles of incidence are progressively refracted downwards and speeded up as they pass through successively lower refractive index values, until they eventually arrive at a point where their angle of incidence is so shallow that total internal reflection occurs and the ray starts to travel back downwards again at ever increasing refractive angles and decreasing speeds, thus following the waveform pattern of a sinewave. This process continuously repeats as the ray propagates through the length of the cable.

The main point to note about the type of photonic wave-travel action described above is that it makes *all* photons travel through the cable at virtually identical mean linear speeds, irrespective of their angles of incidence at the cable's butt-end input. Thus, GRIN cable offers a very low modal dispersion value (typically 0.5ns/km), combined with fairly wide butt-end input acceptance angles and great ease of use. The most popular version of the cable is known by its trade name of 'Widoshi' cable and has fibre dimensions of 62.5/125µm. Typical bandwidth–length product values are 500MHz/km to 1GHz/km; attenuation values of less than 0.22dB/km are attainable under optimum operating conditions. GRIN cables are now widely used in short- to medium-haul high-data-rate commercial applications.

Fibre optic cable summary. The table shown in *Figure 10.11* summarizes all of the above data. Note that polymer 'light pipe' cables are relatively 'big' multi-mode step-index devices that are quite easy to use and are thus popular with amateurs, experimenters, and occasional users of fibre optic cables, but

Fibre optic cable type	Fibre dimensions	Operating wavelengths	Maximum operating bandwidth	Maximum data rate (bits/sec/km)	Basic description
Polymer multi-mode 'light pipe' cable	1mm/2.2mm	400 to 700nm (visible light range)	1MHz/km	500kb/s/km	Easy-to-use cable, for use in very-short-haul very-low-data-rate applications
Multi-mode step index, glass	50/125µm or 62.5/125µm	850nm 1300nm 1550nm	33MHz/km	16Mb/s/km	Used mainly in short- to medium-haul low-data-rate applications
Graded-index (GRIN), glass	50/125µm or 62.5/125µm	850nm 1300nm 1550nm	0.5 to 1.0GHz/km	250Mb/s/km to 500Mb/s/km	Used mainly in short- to medium-haul high-data-rate applications
Single-mode step-index, glass	50/125µm or 62.5/125µm	850nm 1300nm 1550nm	Up to 10GHz/km	Up to 5Gb/s/km	Used mainly in medium- to long-haul very-high-data-rate applications

Figure 10.11 *Table listing the main features of the four major types of fibre optic cable*

have very severe operational (bandwidth and attenuation) limitations. For most high-performance professional uses, glass multi-mode step-index or graded-index (GRIN) cables should be used. Single-mode step-index glass cables are very specialized devices, intended mainly for use in commercial very-high-performance medium- to long-haul applications.

Using fibre optic cable

Figure 10.12 shows the basic elements that are used in a simple 1-way fibre optic communication system. The transmitter (Tx) unit houses a pulse-driven light generator such as a LED or laser diode, which has its output fed into the butt-end input of the cable via a dedicated low-loss coupler. At the other end of the cable, another low-loss coupler feeds the optical signal into a

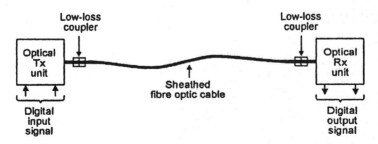

Figure 10.12 *Basic elements of a simple 1–way fibre optic communication system*

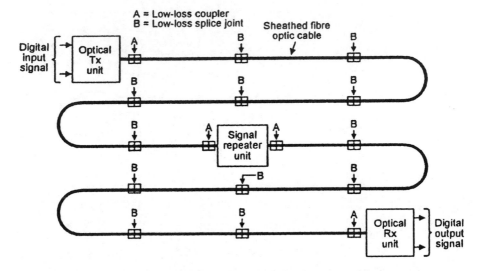

Figure 10.13   *Basic elements of a long 1–way fibre optic communication system using splice joints and signal repeaters*

receiver (Rx) unit that houses a light sensor such as a PIN diode or photo-transistor, plus appropriate signal conditioning circuitry.

It is important to note that the fibre optic communication system will only work efficiently if its cable ends are accurately cut and polished, to minimize attenuation losses, and if special low-loss signal couplers are used. Polymer cable is fairly easy to use, and can be cut reasonably well with a very sharp blade. Glass cable must be cut and polished with the aid of special (and fairly expensive) equipment. A well made coupling gives an attenuation loss of less than 0.5dB; a badly made coupling may give very high attenuation or a complete failure to pass optical signals.

Figure 10.13 shows a long and more complex 1-way fibre optic system that uses splice joints and signal repeaters. Splice joints are used to cleanly join individual cables together, either in the initial cable-laying stages or after making a cable repair, and give a typical coupling loss of 0.5dB per joint. Fibre optic cables are supplied on reels; polymer cables have reel lengths of 20 metres; glass multi-mode step-index and GRIN cables have reel lengths of 100 or 500 metres. Thus, a 5km length of GRIN cable typically uses two input/output couplers plus nine splice joints, which give a combined total attenuation loss of 10.5dB, plus that of the actual cable.

Optical signal repeaters are DC-powered and contain an optical Rx input stage, a signal conditioner, and an optical Tx output stage. When inserted into a fibre optic cable system, they can boost a weak and degraded digital input signal and convert it into a fully restored strong and clean output signal.

Signal repeaters can be used purely as signal boosters, or as signal data-rate (or frequency bandwidth) expanders.

Most fibre optic cable systems are driven by LED or laser diode inputs at power levels in the typical range 0.5mW to 5mW, and when repeaters are used purely as signal boosters in such systems they should be spaced at source-to-repeater 'attenuation' intervals of 30dB maximum. In the case of a GRIN system made from 500 metre lengths of cable with an attenuation value of 1.5dB/km, for example, repeaters should be placed at maximum intervals of 10km (25.5dB attenuation) to 12km (30.5dB attenuation).

Note in the above example that if the GRIN cable has a maximum operating data rate of 250 Mbits/s/km, it will be able to operate at a maximum data rate of 500 Mbits/s in a 0.5km cable, but at only 25 Mbits/s in a 10km cable, and so on. This length-related loss of data rate (and signal bandwidth) can be minimized with the aid of signal repeaters. If these are spaced at 1km intervals along the cable, the data rate will (in this example) not fall below 250 Mbits/sec.

Finally, note that the types of fibre optic cable used in most modern commercial long-haul bidirectional telephone systems are armoured and accommodate up to 35 fibre pairs. Each digitized telephonic speech channel operates at a 70–80 kbits/s data rate, and a cable with a total bit-rate capacity (divided among a number of fibre pairs) of 650 Mbits/s can thus accommodate a total of 8000 individual speech channels.

LEDs and lasers

Introduction

The two basic types of light-generating device most widely used in modern optoelectronics are the LED and the laser (the word LASER is an acronym for *L*ight *A*mplification by *S*timulated *E*mission of *R*adiation). LEDs and lasers differ in two major respects. First, the LED is a semiconductor device that emits light as a consequence of a current-in–photons-out power conversion process, whereas the laser is a tuned-cavity resonator device that may use a gas, liquid, or solid substance as its active medium and emits light as a result of a photon multiplication process.

The second major difference between the two types of device is that the LED emits broad-band light in which its photons are randomly generated and are not directly phase-related, whereas the laser emits a stimulated narrow-band *coherent* type of light in which its emitted photons are – *at the moment of their birth* – of the same wavelength and phase as their parent photons. Because of its coherence, the laser light can be focused into a far smaller spot than that of a LED, thus enabling it to generate very high local power densities. A perfect 0.5mW IR laser beam can, for example, be focused into a minute spot measuring only 1.6μm in width (roughly equal to 1/30th of the width of a human hair), in which the IR power density has a value of about 12kW/cm² within the focused area.

To begin to understand the basic light-generating principles of LEDs and lasers, it is necessary to first learn some basic facts about the nature of atoms and about photon generation.

Atom and photon basics

All solids, liquids and gases are composed of chemical elements, of which there are 109 different known kinds. Each of these elements are made up from clusters, chains or lattices of atoms, and the atom is thus the basic building block of all matter.

All atoms take the basic form illustrated in *Figure 10.14* and consist of a positively charged central nucleus (made up of protons and neutrons) that is surrounded by a number of *bands* of orbiting negatively charged electrons. Normally, the positive nucleus charge and the negative electron charges balance one another, thus giving the atom a neutral overall charge; the atom is said to be *stable* under this condition.

Within the atom, each orbiting electron has a finite kinetic energy value, which determines the distance of its band's orbit from the central nucleus; the individual electron bands are thus known as *energy bands*; their energy is measured in electron-volt (eV) units. Electrons orbiting in energy bands close to the central nucleus have lower energy values than those in the outer bands. The atom's outer electron-occupied energy band is known as the *valence* band.

Beyond the atom's valence band lays a normally empty high-energy *conduction* band. If an electron that is orbiting in the valence band gains

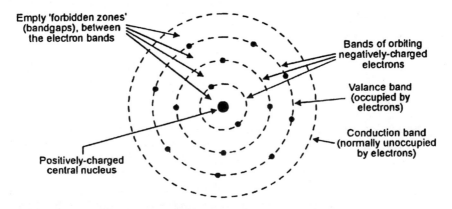

Figure 10.14   *Basic form of an atom*

enough extra energy, it moves upwards into the conduction band, but as it does so it leaves behind a positively charged *hole* in the valance band; the energy-gain process thus creates an electron–hole pair. The atom is said to be in a *quisi-stable* or *excited* state under this condition. When the atom is in this excited state, an electron will readily move downwards from the conduction band into the valence band if it loses sufficient energy, and anihilates the valance band's positively charged hole in the process.

In a piece of matter composed of many atoms, the atoms may be linked together in a matrix-like way via their individual conduction bands. In this case, electrons that are not tied to specific atoms (*free* electrons) may travel freely through the material via the linked conduction bands.

The spaces between the atom's energy bands are known as *forbidden* zones or *bandgaps*. Electrons cannot exist in forbidden zones, but can – if subjected to sufficient energy change – jump through them to reach an adjacent energy band. When an electron jumps downwards through a bandgap (from the conduction band to the valance band) it loses an amount of energy equal to the difference between the two energy band eV values, and this loss is accompanied by the emission of a sub-atomic particle (such as a phonon or photon) with an identical energy value.

From the optoelectronics point of view, all useful photon and electron activity takes place in the vicinity of the atom's valence and conduction bands, and all such activity can thus be represented by simple energy-level diagrams such as those shown in *Figures 10.15* and *10.16*, each of which depicts a narrow 'slice' section taken through the two bands.

Figure 10.15(a) shows the basic energy-level diagram of an atom of a conductive material such as copper. Here, the forbidden zone is so narrow that the valance and conduction bands almost merge into one another, and electrons can – under the influence of an external potential – easily move into the conduction band from the valence band, in which case the vacated valance position is filled by a positively charged hole, as shown in *Figure 10.15(b)*. Once the electron enters the conduction band it is no longer bound

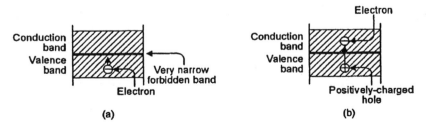

Figure 10.15 *Energy-level diagrams relating to a conductor such as copper. (a) An electron can easily move upwards from the valence band, but (b) leaves a positively charged hole behind*

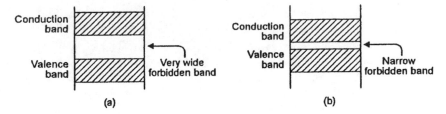

Figure 10.16   *Energy-level diagrams relating to (a) a typical electrical insulator and (b) a typical semiconductor*

to the atom, and is free to travel through the conductive material (via the linked outer bands of adjacent atoms) as an electric current.

Figure 10.16(a) shows the energy-level diagram of an atom of a typical insulation material. Here, the forbidden zone is very wide, thus blocking the flow of electrons into the conduction band and preventing the flow of current through the material. Finally, *Figure 10.16(b)* shows the energy-level diagram of an intrinsic (naturally occurring) semiconductor material such as silicon, which has a narrow forbidden zone and thus has conduction characteristics mid-way between those of a conductor and an insulator. Unlike a normal conductor, however, the semiconductor's resistance has a negative (rather than positive) temperature coefficient.

LED operating principles

Conventional LEDs work in the same basic way as normal silicon junction diodes, but use special semiconductor materials to produce the light-emitting diode's photon-generating characteristics.

Silicon junction diodes are based on two different extrinsic (artificially modified) types of crystalline silicon; one type is very lightly doped with a material such as phosphorus, which has the effect of adding a number of spare electrons (*donors*) to the conduction band of the silicon's crystal lattice, as shown in the material's energy-level diagram of *Figure 10.17(a)*; this material is known as an *n*-type semiconductor, since it carries an excess

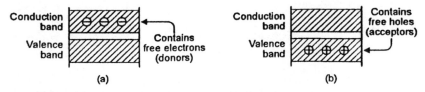

Figure 10.17   *Energy-level diagrams for (a) n-type and (b) p-type semiconductors*

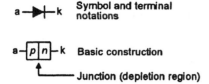

a —▶︎|— k   Symbol and terminal notations

a —| p | n |— k   Basic construction

↑ —— Junction (depletion region)

Figure 10.18 *Symbol and basic construction of a silicon junction diode*

negative charge. The other type of silicon is lightly doped with boron, which has the effect of adding a number of spare holes (*acceptors*) to the valance band of the silicon lattice, as shown in the energy-level diagram of *Figure 10.17(b)*; this material is known as a *p*-type semiconductor, since it carries an excess positive charge.

In a junction diode, the *n*-type and *p*-type materials are fused together in the manner indicated in the greatly simplified diagram of *Figure 10.18*, which also shows the diode's circuit symbol. Since both halves of the diode are made from the same basic material (silicon), the device is often called a *homojunction* diode. *Figure 10.19* shows the diode's energy-level diagrams under *(a)* unbiased, *(b)* reverse biased, and *(c)* forward-biased conditions. In these diagrams, the junction between the *n*-type and *p*-type materials is named the *depletion* region.

The energy level of the junction diode's positively charged *p*-type material is inherently higher than that of the negatively charged *n*-type material, as shown in *Figure 10.19(a)*. Consequently, if the diode is reverse biased as shown in *Figure 10.19(b)*, the energy level difference between the *p*-type and *n*-type material becomes even greater, inhibiting any significant flow of electrons or holes between the two materials via the depletion region. Under this reverse-biased condition, the only currents that flow through the diode are small temperature-sensitive leakage currents and (if the junction is directly exposed to an external light source) small photon-induced currents.

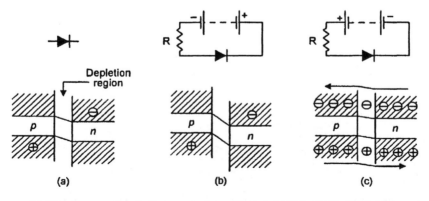

Figure 10.19 *Energy-level diagrams of a silicon junction diode under (a) unbiased, (b) reverse-biased, and (c) forward-biased conditions*

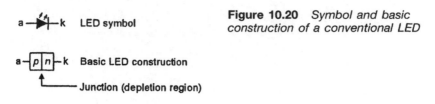

Figure 10.20   *Symbol and basic construction of a conventional LED*

Alternatively, if the diode is forward biased as shown in *Figure 10.19(c)*, the energy level difference between the *p*-type and *n*-type material falls to near-zero at a forward bias value of about 0.6V, enabling electrons and holes to flow freely between the two materials via the depletion region. An electric current thus flows through the forward-biased junction; it actually flows through the covalance bands (valance bands that are linked in adjacent atoms) within the silicon matrix and often generates a phonon particle at the moment of exchange; the phonon energy is dissipated within the crystal lattice as vibrant heat.

Conventional LEDs use the same basic homojunction form of construction and work in the same basic way as normal junction diodes, but use special semiconductor materials (rather than silicon) that emit photons (rather than phonons) when forward current flows through the material's lattice. *Figure 10.20* shows the circuit symbol and the basic construction of a conventional LED, and *Figure 10.21* shows the LED's energy-level diagram under forward-biased operating conditions.

In *Figure 10.21*, when current is flowing through the forward-biased junction, most free electrons and free holes travel through the depletion (junction) area in the normal way, but some electrons don't have enough energy to stay in the conduction band and drop down (via the forbidden bandgap zones of individual atoms) into the valance band and inihilate a hole. The electron energy lost in this process is converted into a photon, which is radiated from the LED as a light particle. LED electrical power-in to optical power-out conversion efficiency is low, typically in the range 0.01% to 1.5%, and is greatly influenced by the photon's wavelength.

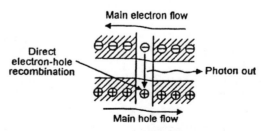

Figure 10.21   *Energy-level diagram of a conventional LED under forward-biased operating conditions*

Semiconductor material	LED colour	Wavelength, λ (nm)	Bandgap energy (eV)
Gallium Nitrogen (GaN)	Blue	430	2.88
Silicon Carbide (SiC)	Blue	480	2.58
Gallium Phosphide (GaP)	Green	565	2.19
Alluminium Gallium Phosphide (AlGaP)	Yellow	595	2.08
Alluminium Gallium Phosphide (AlGaP)	Orange	620	2.00
Alluminium Gallium Arsenide (AlGaAs)	Red	645	1.92
Gallium Alluminium Arsenide (GaAlAs)	Infra-Red	880	1.41
Gallium Arsenide (GaAs)	Infra-Red	950	1.31

Figure 10.22 *Table of LED wavelength/bandgap-energy relationships*

The LED's photon wavelength, λ, (in nm) is dictated by the bandgap energy (eV, the difference between the valance and conduction band energy values) of the LED's semiconductor material, and these two parameters are related by the easily remembered formula $λ = 1240/eV$, and $eV = 1240/λ$. Thus, a LED that generates a red output at a wavelength of 645nm has a bandgap energy value of 1.92 eV. *Figure 10.22* shows some practical LED wavelength/bandgap-energy relationship values, together with basic details of the types of semiconductor material used to make various LEDs. The basic semiconductor material determines the approximate energy value of the bandgap; the actual value is fine-tuned by suitably doping the material.

Note that the bandgap values quoted in *Figure 10.22* are 'mean' (rather than absolute) ones. In reality, the bandgap energy of an atom is not fixed, but varies from moment to moment, depending on the instantaneous depths of individual electrons within its valence–conduction bands. These energy variations are fairly small (usually within ± 2% of mean), but (since a photon's wavelength is directly related to bandgap energy) cause the LED's output to have a finite 'minimum *spectral bandwidth*' value.

The simple LED described earlier in this section is a surface-emitting homojunction type that uses the same basic material on both sides of its junction. In practice, many modern LEDs use different materials on the two junction sides, and are thus known as heterojunction LEDs; most such LEDs have two-stage internal junctions, and are known as double-heterojunction LEDs. Some special LEDs (designed to easily interface with fibre optic cables) are 'edge-emitting' types that emit a narrow beam of light from the side of the semiconductor material, rather than a broad beam from its face.

Laser basics

LEDs emit individual photons in a quite random or *spontaneous* fashion, whereas lasers emit photons in a *stimulated* fashion in which the birth of each new photon occurs on the arrival of a parent photon. In a laser, each new

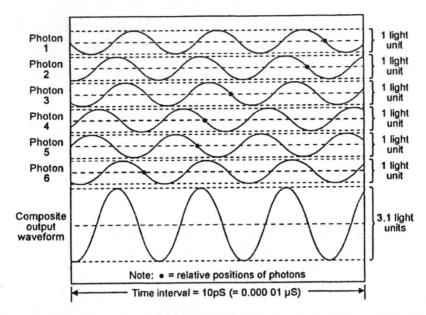

Figure 10.23   *Six typical random (spontaneous) photon waveforms emitted by a 850nm IR LED, together with their composite 'light' waveform*

photon is, at the moment of its birth, a duplicate of its parent photon, with the same basic wavelength and (more importantly) an identical phase. Light composed of in-phase photons is known as *coherent* light, and can be focused far more sharply than other types of light.

Figure 10.23 shows six typical 'random' photon waveforms emitted by a LED, together with their composite light output waveform (equal to the sum of the six photon waveforms), and *Figure 10.24* shows a similar set of coherent waveforms emitted by a laser. Each diagram spans a 10ps (10 picosecond) time interval (1ps = 1 millionth of a μs) and depicts six newly generated IR photons with wavelengths of 850nm (= a frequency of 353GHz) and an amplitude value of '1 light unit'. The typical relative position of each photon is indicated by a small black dot.

Note in *Figure 10.23* that the waveforms of photons 1 and 5 are in exact anti-phase and thus have a combined value of zero, and that all remaining photon waveforms are out of phase with one another and (in this particular case) have a combined value of 3.1 light units. In practice, the pattern of the six photon waveforms changes continuously as new photons enter the time frame from the left and others leave the frame on the right, and this causes the instantaneous phase and amplitude of the composite output waveform to vary over a wide range. As a consequence of these rapid phase and amplitude variations (intermodulation), the LED's light output is very impure and has a fairly wide spectral bandwidth.

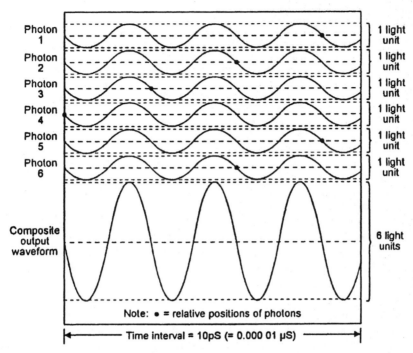

Figure 10.24 *Six typical* coherent *(in-phase) photon waveforms emitted by a 850nm IR laser, together with their composite 'light' waveform*

By contrast, the six laser-generated photon waveforms shown in *Figure 10.24* are all exactly in-phase, and the composite output waveform thus has a fixed phase relationship and has a constant amplitude of 6 light units. The laser's light output is thus very pure, with a very narrow spectral bandwidth. Typically, a modern 850nm IR LED has a half-power spectral bandwidth (i.e. the bandwidth at which the light power output falls to half of its maximum value) of about 80nm. Modern 850nm laser diodes, on the other hand, have typical half-power bandwidths of 0.5nm to 5nm, and their light outputs are thus far purer than those of LEDs.

Laser operating principles

The basic *atomic* operating principle of the laser is in some ways similar to that of the LED. In both cases, a photon is generated in an excited atom when an electron in its conduction band loses energy and drops down into the atom's valence band, annihilating a hole and generating a photon in the process. In the LED, this process occurs randomly, when the electron's energy decays below a critical value. In the laser, however, the process is

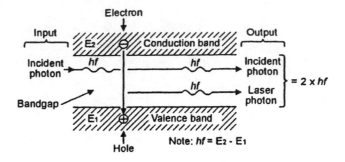

Figure 10.25   *Basic atomic operation of the lasing process*

initiated by photonic stimulation, in which an external photon *with an energy value equal to that of the atom's bandgap* enters the atom's bandgap and, by a quantum process known as negative absorption, makes the electron lose energy and drop down into the atom's valence band, thereby anihilating a hole and generating an identical photon.

Figure 10.25 illustrates the basic atomic operation of the 'lasing' process. Here, the incident (arriving) photon, which has an energy value identical to that of the atom's conduction–valance bandgap (= *hf*), enters the bandgap from the left side of the diagram and makes the electron drop down and annihilate the hole, thereby generating a new 'laser' photon, which also has an energy value of *hf*. The incident and laser photons, which have identical phase and frequency-wavelength values, then emerge together on the right side of the diagram.

Note from the above descriptions that lasing action can only occur if the atom is in an excited state at the moment that the incident photon arrives, and if the incident photon's energy value equals that of the atom's bandgap. Also note that, if lasing action does occur, the excited atom gives an effective 2:1 overall photonic power gain, but drops out of the vital 'excited' state after generating a single photon.

In a practical laser, the actual lasing medium (which contains vast numbers of atoms capable of giving lasing action) may take the form of a gas, a liquid, or a solid material, in which the lasing atoms may be raised into an excited state by electrical, chemical or optical means. In laser jargon, the process of feeding energy into the medium to hold its atoms in an excited state is known as *pumping*, and the situation in which most of the medium's atoms are in the excited state is known as a situation of *population inversion*. Thus, the medium must be pumped to create population inversion, which is a prerequisite of laser action. Lasing action can only begin if the medium's energy input exceeds a certain *pumping threshold* value; to give continuous lasing, the medium must be continuously pumped.

Figure 10.26 shows the basic elements of a complete laser unit. One of these is the actual lasing medium, which must be reasonably translucent.

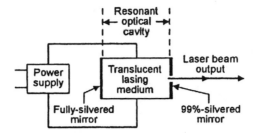

Figure 10.26 *Basic elements of a complete laser unit*

Another is a power supply that is used to pump the medium into a state of population inversion and initiate the lasing action. The final unit is a resonant optical cavity, made of two mirrors set at opposite ends of the lasing medium so that photons can repeatedly bounce back and forth through the medium; one mirror is fully reflective; the other has a small (equal to about 1% of the mirror's surface area) translucent hole in its centre.

In *Figure 10.26*, lasing begins when – with the medium pumped up to its threshold value – a suitable internally- or externally generated incident photon hits an excited atom and initiates the birth of an identical photon. This pair of photons then travel to the right, hit the end of the cavity, then reflect back and forth through the cavity, repeatedly passing through the lasing medium and initiating the generation of more photons in each pass, so that the photon flow rapidly builds up into a flood that can easily be controlled via the power supply. Photons that strike the translucent hole at the centre of the cavity's right-hand mirror emerge (*bleed off*) from the laser unit in a narrow and coherent beam that can be focused into a minute and intense spot by an external lens.

The optically resonant cavity is a vital part of the laser system. It is highly wavelength-selective and gives high optical gain (resonance) only to signals with an integer number of half-wavelengths that fit exactly into the cavity's *optical* length (equal to the product of its physical length and refractive index value). The cavity's length is inevitably many times longer than the (roughly 1000nm) optical wavelength of the basic laser beam, and it thus has a huge number of optical operating *modes* (wavelengths at which resonance may occur). In a cavity with an optical length of 2mm, for example, the 1st mode occurs at a wavelength of 4mm, the 10th at 0.4mm, the 100th at 40µm, the 1000th at 4000nm, the 4705th at 850.16nm, the 4706th at 850nm (a commonly-used IR wavelength), and the 4707th at 849.8nm.

The optical cavity's frequency response thus consists of a series of sharp comb-like peaks (known as *mode lines* in laser jargon), as shown in *Figure 10.27(a)*; in the above example, at wavelengths around 850nm, these peaks are spaced roughly 0.2nm apart. In some practical lasers (including many laser diodes), the lasing medium's basic (unfiltered) output has the type of

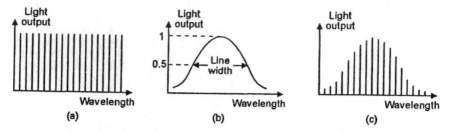

Figure 10.27 *Typical frequency response of (a) a laser's optically resonant cavity, (b) the unfiltered output of a lasing medium, and (c) the resulting laser-beam output of the cavity*

frequency response shown in *Figure 10.27(b)*, and spans dozens of these peaks; in such cases, the cavity's laser-beam output thus has the type of frequency response shown in *Figure 10.27(c)*.

In laser jargon, the half-power spectral bandwidth of a laser output is called its *linewidth*. Lasers with linewidths that fit into a single mode line are known as *single mode* lasers; those with linewidths that span more than one mode line (as in the case of *Figure 10.27(c)*) are known as *multi-mode* lasers. All practical lasers have finite linewidths and (since they contain a spectrum of frequencies) their beams can remain fully coherent (in-phase) for only a limited number of wavelengths after their initial creation. This dimension is known as the beam's *coherence length*, and equals the square of the beam's mid-value wavelength divided by its linewidth. Thus, an 850nm beam with a linewidth of 2nm has a coherence length of only 0.36mm.

Laser types

The world's first working optical laser was demonstrated by an American, Theodore Maiman, in mid-1960. Known as a ruby laser, it has the basic form shown in *Figure 10.28(a)* and consists of a small rod of synthetic ruby crystal (the lasing medium), with mirrored ends that form the optical cavity; the rod is surrounded by an electrically powered helical flash tube, which acts as the medium's energy pump. When the flash tube is operated, its white light pumps the rod's atoms into an excited state, and a brief burst of lasing action commences a few milliseconds later, causing a brief pulse of red laser light to emerge from one end of the tube.

The ruby laser provides only brief pulses of laser light. The first laser to give continuous-wave (cw) operation was the helium–neon (He–Ne) gas laser, which was first demonstrated in late 1960 and has the basic form shown in *Figure 10.28(b)*. Here, the high-voltage supply pumps the gas molecules into an excited conductive state in which sustained lasing action takes place (typical supply voltages are 8kV for starting, 1.5kV when running); lasing

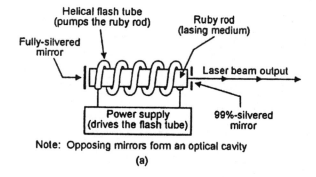

Note: Opposing mirrors form an optical cavity

(a)

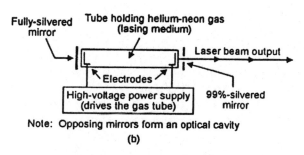

Note: Opposing mirrors form an optical cavity

(b)

Figure 10.28 *Basic functional diagrams of (a) ruby and (b) helium–neon lasers*

takes place primarily at a wavelength of 632nm and the device thus generates a bright red laser beam. He–Ne lasers are still widely used; most have maximum optical output powers in the range 0.5 to 10mW; the outputs have typical linewidths of a mere 0.002nm and beam coherence lengths of 200mm. Some modern He–Ne lasers have outputs that can be electronically modulated (but not pulsed) at frequencies up to 1MHz.

Many other types of laser have appeared since 1960. Most have severe practical disadvantages (such as very brief working lives, excessive cost, fragility or bulkiness, or severe thermal operating requirements) or are meant only for laboratory or military use. Some are designed specifically for use in optical welding or cutting operations, and are not suitable for use in general 'electronic' applications. Most have input-to-output power conversion efficiencies of only 0.001 to 0.5 percent (a 1mW He–Ne laser, for example, typically needs an input power of 20W).

The laser best suited to use in general electronic applications is the so-called 'laser diode', which (in most low-power types) acts like a laser version of the double heterojunction side-emitting LED, with a built-in resonant optical cavity. It is compact, robust, has a typical working life of 50 000+ hours, and is reasonably easy to use.

A laser diode produces a cw laser beam that can be pulse or analogue modulated at frequencies up to many hundreds of megahertz, has a typical

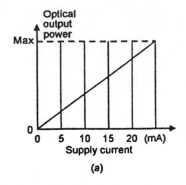
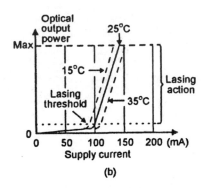

Figure 10.29 *Typical supply-current to optical-output-power graphs of (a) a normal LED and (b) a laser diode*

input-to-output power conversion efficiency of 1.5%, and is commercially available in versions giving maximum optical power outputs ranging from 0.1mW to several watts, at a variety of red and infra-red wavelengths. Single mode types have typical linewidths of less than 0.5nm; multi-mode types have typical linewidths in the range 2nm to 5nm.

Figure 10.29(a) shows the typical supply-current to optical-output-power graph of a normal LED, and *Figure 10.29(b)* shows that of a laser diode. The LED's graph is quite linear, and the optical output power is directly proportional to the LED's supply current value. The laser diode has a more complex performance graph; at supply currents well below the lasing threshold value it acts like a normal LED and generates random photons, but at currents well above the lasing threshold it gives true lasing action and generates coherent photons; when operating at the actual lasing threshold level, it generates both random and coherent photons.

Note in *Figure 10.29(b)* that, at any fixed lasing current value, the optical output power is very sensitive to variations in the laser diode's temperature. Most practical laser diode units (modules) incorporate sensing and control circuitry that stabilizes the optical output power by auto-adjusting the lasing drive current. Most modern units of this type use special laser diodes that have an integral monitoring photodiode; *Figure 10.30* shows a 4-pin version of such a unit, together with a simple way of using it as a cw laser transmitter.

In *Figure 10.30*, the photodiode's reverse current (and thus the voltage on Q1 base) is proportional the laser beam's intensity. Q1 acts as a voltage-controlled constant-current generator that supplies base drive to Q2, which in turn provides drive current to the laser diode. The overall action is such that, if the laser intensity falls below a value preset via RV1, the photodiode-derived Q1–base voltage also falls, thereby increasing Q1's current drive to Q2 base and causing Q2 to increase the drive current to the laser diode, which responds by increasing its laser intensity. The reverse process occurs

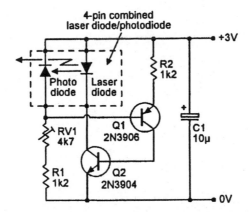

Figure 10.30 *Simple way of using a 4-pin combined laser diode/photodiode unit in a cw laser transmitter*

if the laser intensity rises above the preset value. The circuit thus auto-regulates the laser beam intensity.

Figure 10.31 shows a basic way of using the 4-pin combined laser diode/photodiode unit in a pulse-modulated cw laser transmitter. Here, the pulsed output of the laser-monitoring photodiode is converted into a dc offset voltage via an integrating amplifier, is combined with an RV1-derived pre-set dc bias voltage via an adder, and is fed to one onput of the laser-driving pulsing and biasing circuit, which has its other input driven by external pulse-coded waveforms. The RV1-derived dc bias voltage is set so that – in the absence of input pulses – the laser diode is biased at its lasing threshold value (rather than fully off), thus giving a very fast switching action.

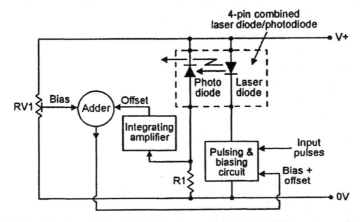

Figure 10.31 *Basic circuit of a pulsed cw laser transmitter using a 4-pin combined laser diode/photodiode unit*

Laser applications

WARNING. Sensible safety precautions must always be taken when using lasers; the output beam of even a 0.5mW type can cause severe optical damage if aimed (deliberately or accidentally) at a human eye.

Low-power (up to 5mW) lasers are readily available in 'module' form, comprising the actual laser plus its drive circuitry and optics. They have a multitude of practical applications. Simple cw types, using the basic type of circuit shown in *Figure 10.30*, can – if generating a visible red beam – be used as laser pointers, spirit levels or alignment aids, or as the basis of bar-code readers. Alternatively, infra-red types can be used in a wide range of security applications, activating an alarm or initiating some other action when a person, animal, object, vehicle, smoke or fog breaks or reflects the beam. When reflected from a solid surface, the cw beam can also be used to detect minute vibrations of the surface.

All CD and CD-ROM players incorporate a small cw laser module that (aided by various lenses and servo-drivers) automatically scans the spinning CD and picks out the information contained in the tracks (which are spaced only 1.6mm apart) of its coded sub-surface data pits, which are each a mere 0.1μm deep and 0.8μm to 3.5μm long, retrieving the data at a rate of over 4 000 000 bits per second.

Laser modules that are designed for use as modulated cw types can be used in all of the types of 'security' applications already mentioned, plus various speed- and distance-measuring application. They are particularly useful in long-haul fibre optic communication and data-transfer systems, in which (because of their very small spectral bandwidths) they can greatly outperform LEDs. Single-mode IR laser diodes, for example, have spectral linewidths of only 0.5nm, compared to 80nm for LEDs, and their signals thus suffer far less material dispersion than those of LEDs when travelling through long lengths of fibre optic cable.

11

Light-beam alarms

In the Chapter 1 survey of modern optoelectronic devices and techniques brief mention was made of light-beam systems that can (among other things) be used as the basis of intrusion detectors and alarms. The present chapter expands on this theme by showing practical ways of making invisible IR light-beam alarms; the chapter starts off by looking at some basic principles.

Intrusion alarm basics

A simple invisible-light-beam intrusion detector or alarm system can be made by connecting an infra-red (IR) light transmitter and receiver as shown in *Figure 11.1*. Here, the transmitter feeds a coded signal (often a simple squarewave) into an IR LED which has its output focused (via a moulded-in lens in the LED casing) into a fairly narrow beam that is aimed at a matching IR photodetector (phototransistor or photodiode) in the remotely placed receiver. The system action is such that the receiver output is off when the light-beam reaches the receiver, but turns on and activates an external alarm, counter, or relay if the beam is interrupted by a person, animal, or object. This basic type of system can be designed to give an effective detection range

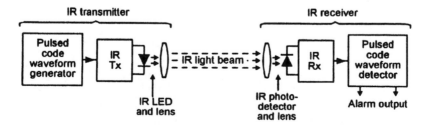

Figure 11.1 *Simple light-beam intrusion alarm/detector system*

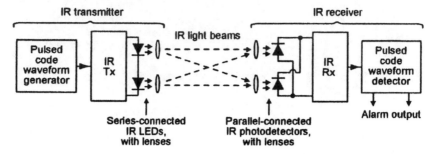

Figure 11.2 *Dual-light-beam intrusion alarm/detector system*

of up to 30 metres when used with additional optical focusing lenses, or up to 8 metres without extra lenses.

The above system works on the pin-point line-of-sight principle and can be activated by any 'wider-than-a-lens' object that enters the line-of-sight between the transmitter and receiver lenses. Thus, a weakness of this simple system is that it can be false-triggered by a fly or moth (etc.) entering the beam or landing on one of the lenses. The improved dual-light-beam system of *Figure 11.2* does not suffer from this defect.

The *Figure 11.2* system is basically similar to that already described, but transmits the IR beam via two series-connected LEDs that are spaced about 75mm apart, and receives the beam via two parallel-connected photodetectors that are also spaced about 75mm apart. Thus, each photodetector can detect the beam from either LED, and the receiver thus activates only if BOTH beams are broken simultaneously, and this normally occurs only if a large (greater than 75mm) object is placed within the composite beam. This system is thus virtually immune to false triggering by moths, etc.

Note that, as well as giving excellent false-alarm immunity, the dual-light-beam system also gives (at any given LED drive-current value) double the effective detection range of the simple single-beam system (i.e. up to 16 metres without additional lenses), since it has twice as much effective infra-red transmitter output power and twice the receiver sensitivity.

System waveforms

Infra-red beam systems are usually used in conditions in which high levels of ambient or background IR radiation (usually generated by heat sources such as radiators, tungsten lamps, and human bodies, etc.) already exist. To enable the systems to differentiate this background radiation and give good effective detection ranges, the transmitter beams are usually frequency modulated, and the receivers are fitted with matching frequency detectors.

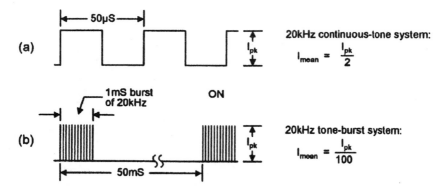

Figure 11.3 *Alternative types of IR light-beam code waveforms, with typical parameter values*

In most cases, the transmitted beams use either continuous-tone or tone-burst frequency modulation, as shown in *Figure 11.3*.

Infra-red LEDs and photodetectors are fast acting devices, and the effective range of an IR beam system is thus determined by the *peak* current fed into the transmitting LED, rather than by its mean LED current. Thus, if the waveforms of *Figure 11.3* are used in transmitters giving peak LED currents of 100mA, both systems give the same effective operating ranges, but the *Figure 11.3(a)* continuous-tone transmitter consumes a mean LED current of 50mA, while the tone-burst system of *Figure 11.3(b)* consumes a mean current of only 1mA (but requires more complex circuit design).

The operating parameters of the tone-burst waveform system require some consideration, since the system actually works on the 'sampling' principle. For example, it is a fact that at normal walking speed a human takes about 200ms to pass any given point, so a practical IR light-beam burglar alarm system does not need to be turned on continuously, but only needs to be turned on for brief 'sample' periods at repetition periods far less than 200ms (at, say, 50ms); the sample period should be short relative to repetition time, but long relative to the period of the tone frequency. Thus, a good compromise is to use a 20kHz tone with a burst or sample period of 1ms and a repetition time of 50ms, as shown in *Figure 11.3(b)*.

System design

The first step in designing any electronics system is that of drawing up suitable block diagrams; *Figure 11.4* shows a suitable block diagram of a continuous-tone IR intrusion alarm/detector system, and *Figure 11.5* shows that of a tone-burst system. Note that a number of blocks (such as the IR output stage, the tone pre-amp, and the output driver) are common to both systems.

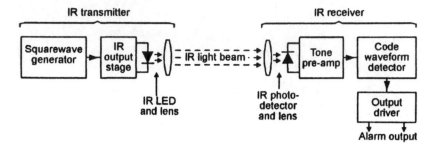

Figure 11.4 *Block diagram of continuous-tone IR light-beam intruder alarm/detector system*

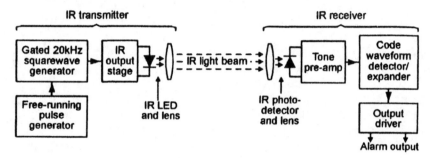

Figure 11.5 *Block diagram of tone-burst IR light-beam intruder alarm/detector system*

The continuous-tone system (*Figure 11.4*) is very simple, with the transmitter comprising nothing more than a squarewave generator driving an IR output stage, and the receiver comprising a matching tone pre-amplifier and code waveform detector, followed by an output driver stage that activates devices such as relays and alarms, etc.

The tone-burst system is far more complex, with the transmitter comprising a free-running pulse generator (generating 1ms pulses at 50ms intervals) that drives a gated 20kHz squarewave generator, which in turn drives the IR output stage, which finally generates the tone-burst IR light-beam. In the receiver, the beam signals are picked up and passed through a matching pre-amplifier, and are then passed on to a code waveform detector/expander block, which ensures that the alarm does not activate during the 'blank' parts of the IR waveform. The output of the expander stage is fed to the output driver.

Figure 11.6 shows an alternative version of the tone-burst system. This is similar to the above, except that a simple code waveform detector is used in the receiver section, and that a blanking gate is interposed between the detector and the output driver and is driven directly by the transmitter's

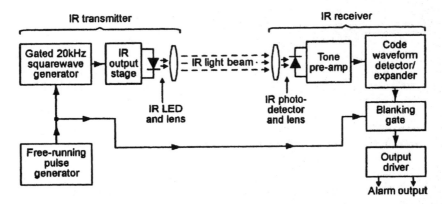

Figure 11.6 *Block diagram of alternative IR light-beam system*

pulse generator, to ensure that the alarm is not activated during the blank parts of the IR waveform.

Transmitter circuits

Figure 11.7 shows the practical circuit of a simple continuous-tone dual-light-beam IR transmitter. Here, a standard 555 timer IC is wired as an astable multivibrator that generates a non-symmetrical 20kHz squarewave output that drives the two IR LEDs at peak output currents of about 400mA via R4 and Q1 and the low source impedance of storage capacitor C1. The timing action of this circuit is such that the ON period of the LEDs is controlled by C2 and R2, and the OFF period by C2 and (R1 + R2), i.e. so that the LEDs are ON for only about one eighth of each cycle; the circuit thus consumes a *mean* current of about 50mA.

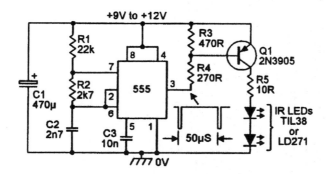

Figure 11.7 *Simple continuous-tone IR light-beam transmitter*

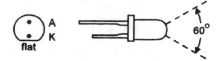

Figure 11.8   *Outline and connections
used by the LD271 and TIL38 IR LEDs*

The above circuit can use either TIL38 or LD271 (or similar) high-power
IR LEDs. These popular and widely available LEDs can handle mean
currents up to only 100mA or so, but can handle brief repetitive peak
currents several times greater than this value. *Figure 11.8* shows the outline
and connections of these devices, which have a moulded-in lens that focuses
the output into a radiating beam of about 60° width; at the edges of this beam
the IR signal strength is half of that at the centre of the beam.

Minor weaknesses of the IR output stage (Q1 and R3–R4) of the *Figure
11.7* circuit are that it has a very low input impedance (about 300 ohms), that
it gives an inverting action (the LEDs are ON when the input is low), and
that the LED output current varies with the circuit supply voltage. *Figure
11.9* shows an alternative universal IR transmitter output stage that suffers
from none of these weaknesses.

In *Figure 11.9*, the base drive current of output transistor Q2 is derived
from the collector of Q1, which has an input impedance of about 5k0 (determined mainly by the R1 value). Thus, when the input is low Q1 is off, so Q2
and the two IR LEDs are also off, but when the input is high Q1 is driven
to saturation via R3, thus driving LED1 (a standard red LED) and Q2 and
the two IR LEDs on.

Note that under this latter condition about 1.8V is developed across LED1,
and that about 0.6V less than this (= 1.2V) is thus developed across R4.
Consequently, since the R4 voltage is determined by the Q2 emitter current,
and the Q2 emitter and collector currents are virtually identical, it can be
seen that Q2 acts as a constant-current generator under this condition, and

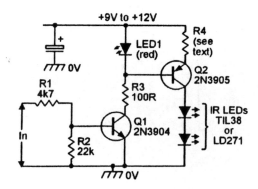

Figure 11.9   *'Universal' IR transmitter output stage*

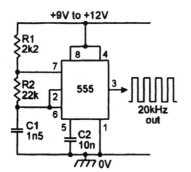

Figure 11.10 *20kHz squarewave 'tone' generator*

that the IR LED drive currents are virtually independent of variations in supply voltage. The peak LED drive current thus equals approximately 1.2/R4, and R4 (in ohms) = 1.2/I, where I is the peak LED current in amps.

Figure 11.10 shows a 20kHz squarewave generator (made from a 555 timer IC) that can be added to the *Figure 11.9* output stage, to make a continuous-tone IR beam transmitter. In this case R4 (in *Figure 11.9*) should be given a value of at least 6.8 ohms, to limit peak LED currents to less than 200mA.

Alternatively, *Figure 11.11* shows the circuit of a tone-burst generator (giving 1ms bursts of 20kHz at 50ms intervals) that can be added to *Figure 11.9* to make a tone-burst transmitter. Here, two sections of a 4011B CMOS quad 2-input NAND gate IC are wired as a non-symmetrical astable multi-vibrator producing 1ms and 49ms periods; this waveform is buffered by a third 4011B stage and used to gate a 20kHz '555' astable via D2, and the output of the 555 astable is then inverted via a fourth 4011B stage, ready for feeding to the transmitter output stage.

Note when using the *Figure 11.11* circuit that R4 in *Figure 11.9* can be given a value as low as 2.2 ohms, to give peak output currents of about 550mA, but that under this condition the transmitter consumes a mean

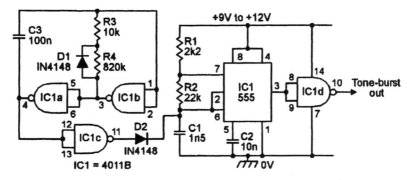

Figure 11.11 *Tone-burst (1µs burst of 20kHz at 50ms intervals) waveform generator*

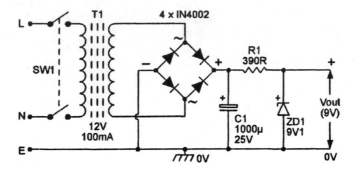

Figure 11.12   *Mains-powered 9V DC supply*

current of little more that 6mA; this current can be provided by either a battery or a mains-derived supply; a suitable mains-powered 9V DC supply is shown in *Figure 11.12*.

A Receiver pre-amp

Figure 11.13 shows the practical circuit of a high-gain 20kHz tone pre-amplifier designed for use in an infra-red receiver. Here, the two IR detectors are connected in parallel and wired in series with R1, so that the detected IR signal is developed across this resistor. This signal is amplified by cascaded op-amps IC1 and IC2, which can provide a maximum signal gain of about ×17 680 (= ×83 via IC1 and ×213 via IC2), but have their gain fully variable via RV1. These two amplifier stages have their responses centred on 20kHz, with 3rd-order low-frequency roll-off provided via C4–C5–C6, and 3rd-order high-frequency roll-off provided by C3 and the internal capacitors of the ICs.

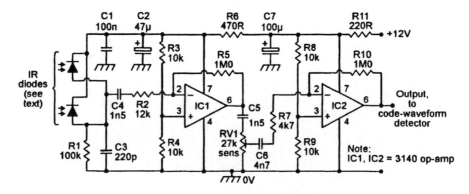

Figure 11.13   *Infra-red receiver pre-amplifier circuit*

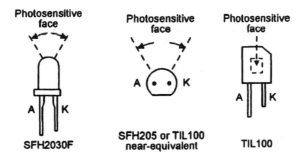

Figure 11.14 *Case outline and IR-sensitive face positions of three popular types of IR photodiode*

The above circuit can be used with a variety of IR detector diode types, which ideally should be housed in black (rather than clear) infra-red transmissive mouldings, which greatly reduce ambient white-light interference. *Figure 11.14* shows the case outline and IR-sensitive face positions of three popular IR photodiodes of this type.

The output of the *Figure 11.13* pre-amplifier can be taken from IC2 and fed directly to a suitable code-waveform detector circuit, such as that shown in Figure *11.15*. Note, however, that if the Tx–Rx light-beam system is to be used over ranges less than 2 metres or so the pre-amp output can be taken directly from IC1, and all the RV1 and IC2 circuitry can be omitted from the pre-amp design.

Code waveform detector

In the *Figure 11.15* code waveform detector circuit the 20kHz tone waveforms (from the pre-amp output) are converted into dc via the

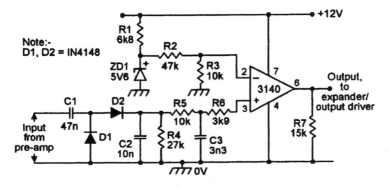

Figure 11.15 *Code waveform detector circuit*

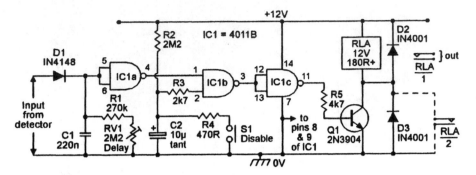

Figure 11.16   *Expander/output driver circuit*

D1–D2–C2–R5–C3 network and fed (via R6) to the non-inverting input of the 3140 op-amp voltage comparator, which has its inverting input connected to a thermally stable 1V0 dc reference point. The overall circuit action is such that the op-amp output is high (at almost full positive supply rail voltage) when a 20kHz tone input signal is present, and is low (at near-zero volts) when a tone input signal is absent; if the input signal is derived from a tone-burst system, the output follows the pulse-modulation envelope of the original transmitter signal. The detector output can be made to activate a relay in the absence of a beam signal by using the expander/output driver circuit of *Figure 11.16*.

Expander/output driver

The operating theory of the *Figure 11.16* circuit is fairly simple. When the input signal from the detector circuit switches high C1 charges rapidly via D1, but when the input switches low C1 discharges slowly via R1 and RV1; C1 thus provides a dc output voltage that is a 'time-expanded' version (with expansion pre-settable via RV1) of the dc input voltage. This dc output voltage is buffered and inverted via IC1a and used to activate relay RLA via Q1 and an AND gate made from IC1b and IC1c.

Normally, the other (pin-2) input of this AND gate is biased high via R2, and the circuit action is such that (when used in a complete IR light-beam system) the relay is off when the beam is present, but is driven on when the beam is absent for more than 100ms or so. This action does not occur, however, when pin-2 of the AND gate is pulled low; under this condition the relay is effectively disabled.

The purpose of the R2–C2 network is to automatically disable the relay network via the AND gate (in the way described above) for several seconds after power is initially connected to the circuit or after DISABLE switch S1

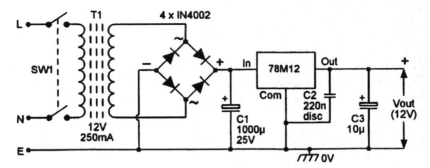

Figure 11.17 *Mains-powered regulated 12V DC supply*

is briefly operated, so that the owner or other authorized persons can safely pass through the beam without activating the relay. Note that the relay can be made self-latching, if required, by wiring normally open relay contacts RLA/2 between Q1 emitter and collector, as shown dotted in the diagram.

A power supply

The circuits of *Figures 11.13, 11.15* and *11.16* can be directly interconnected to make a complete infra-red light-beam receiver that can respond to either continuous-tone or tone-burst signals. Such a receiver should be powered via a regulated 12V DC supply; *Figure 11.17* shows the circuit of a suitable mains-powered unit.

12

IR remote-control systems

Chapter 1 of this volume briefly mentioned the fact that IR light-beam systems can be used for remote-control purposes. This final chapter of the book expands on this theme by explaining the basic operating principles of IR remote-control systems, and by giving hints on how to use modern IR remote-control ICs.

Remote-control systems

All modern remote-control systems take the basic form shown in *Figure 12.1*, and consist of a remote-control transmitter and a remote-control receiver that are linked by some type of transmission medium. At the transmitter end of the system, a 'control' instruction is selected via the keyboard, is converted into a multi-bit digital word by a digital encoder, and is then transmitted *in serial form* by the transmitter's Tx unit. At the receiver end of the system the transmission medium's coded signal is picked up by the Rx unit, is decoded by a digital decoder, and is then converted into some useful form by an 'output actions' unit.

The 'transmission medium' mentioned in the above paragraph may take a variety of forms, including a direct wire or fibre optic cable link, a wireless link, or an IR light-beam link. Wireless links offer good effective operating

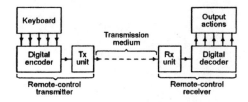

Figure 12.1 *Basic form of a modern remote-control system*

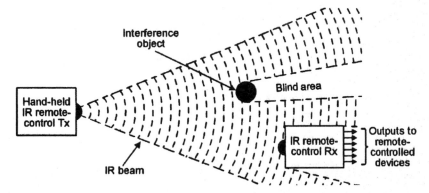

Figure 12.2 *Diagram illustrating some important basic principles of IR remote-control operation*

ranges and are not obstructed by walls or various other physical structures, but (in 'security' applications) can easily be intercepted and duplicated by criminals equipped with scanner/recorder units. IR light-beam links work on a strict line-of-sight principle and have fairly short operating ranges (typically less than 12 metres), but have good immunity to interception/duplication by criminals. The rest of this chapter is concerned primarily with remote-control systems that use IR light-beam links.

Figure 12.2 illustrates some important basic principles of IR remote-control operation. Here, the hand-held control unit transmits a coded waveform via a broad IR light beam, and this signal is detected and decoded in the remotely placed receiver and thence used to activate external devices, etc., via the receiver outputs. Note that the transmitter can remote-control a receiver that is placed anywhere within the active area of the IR beam, and that the transmitter and receiver do not need to be pointed directly at each other to effect operation but *must* be in line-of-sight contact; also note that an object placed within the beam can create a blind area in which line-of-sight contact cannot exist.

Code waveforms

TV/Hi-Fi controllers

Most modern IR remote-control systems give multichannel operation, with each channel giving digital control of an individual function. The transmitter waveforms usually take the general form shown in *Figure 12.3*, which depicts those of a basic 6-bit multichannel system. Here, the waveform comprises a 9ms repeating frame of seven bits of pulse-coded information, with each bit modulated at about 30kHz. The first bit has a fixed 1ms duration and

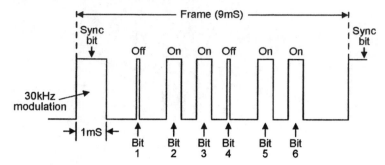

Figure 12.3   *Typical transmitter code waveform of a 6–bit remote-control system*

Channel number	Channel state	Decoded function
1	On Off	Switch A On Switch A Off
2	On Off	Switch B On Switch B Off
3	On Off	Volume increase
4	On Off	Volume decrease
5	On Off	Brilliance increase
6	On Off	Brilliance decrease

Figure 12.4   *Typical functions of a 6–channel simultaneous remote-control system*

provides frame synchronization for the decoder; the subsequent six data bits appear at 1ms intervals and each gives an on/off form of control, with a less-than-0.25ms pulse representing an OFF or logic-0 state and a greater-than-0.25ms pulse representing an ON or logic-1 state. In practice, this 6-bit code waveform can be used to give either six simultaneous channels or up to 64 non-simultaneous channels of remote-control, as shown in *Figures 12.4* and *12.5*.

When the *Figure 12.3* waveforms are used to give simultaneous operation, each data bit controls a single channel, as shown in *Figure 12.4*. Channels 1 and 2 may thus each be used to give independent on/off switching functions, while channels 3 and 4 may be used to increase or decrease the output of a ramping volume control, and channels 5 and 6 may be used to similarly control the output of a ramping brilliance control, etc. Note that, within each transmitter frame, all six channels can be varied simultaneously.

Channel number	6-Bit code	Decoded function	
1	000 000	Switch A	On
2	000 001	Switch A	Off
3	000 010	Switch B	On
4	000 011	Switch B	Off
61	111 100	Volume increase	
62	111 101	Volume decrease	
63	111 110	Brilliance increase	
64	111 111	Brilliance decrease	

Figure 12.5 *Typical functions of a 64–channel non-simultaneous remote-control system*

When the *Figure 12.3* waveforms are used to give non-simultaneous remote-control operation the six data bits are used (within each frame) to generate a unique 6-bit binary code, and each of these codes controls an individual remote-control channel; there are a total of 64 possible codes, and this system can thus be used to give up to 64 channels of remote-control, as shown in *Figure 12.5*. Thus, channels 1 to 4 may be used to control the on/off action of a pair of switches, and channels 61 to 64 may be used to control the levels of volume and brilliance; the remaining 56 channels may be used for a variety of other purposes. Note that, within each transmitter frame, only a single channel can be controlled at a time but that, since the frames are updated several times per second, this is usually only a minor disadvantage.

The 6-bit type of binary-coded waveform shown in *Figure 12.3* can produce a maximum of 64 different binary combinations. This number is adequate for use in most remote-control TV and Hi-Fi applications, but is not adequate for use in security applications such as remote-controlled locks and burglar alarms, etc., where many thousands of different possible combinations are needed for reasonable security. In the latter cases, systems using a dozen or more bits per frame may be needed.

'Security' controllers

Most modern multi-bit remote-controller systems use normal binary coding, in which each bit has two possible states ('high' or 'low'). Some modern CMOS-based systems, however, use trinary or '3–state' coding, in which each bit has three possible states (high, low, or open-circuit). The maximum number of possible bit combinations equals 2^N in binary systems and 3^N in trinary systems (where N = the number of bits per frame). The *Figure 12.6* table translates this data into practical figures. Thus, a 14-bit binary system offers 16 384 possible code combinations, whereas a 9-bit trinary system (such as that used in the SGS Thomson M145026/7/8 series of CMOS-type remote-control ICs) offers 19 683 possible code combinations.

Number of bits (= N)	Maximum possible 'bit' combinations in:-	
	Binary code $(= 2^N)$	Trinary code $(= 3^N)$
4	16	81
5	32	243
6	64	729
7	128	2187
8	256	6561
9	512	19,683
10	1024	59,049
12	4096	531,441
14	16,384	4,782,969
16	65,536	43,046,721

Figure 12.6   *Maximum possible 'bit' combinations available from various code-word formats*

To put the above data into perspective in security applications, note that a mechanical key-operated lock's equivalent of 'possible code combinations' is its number of 'differs' or possible key profiles; Yale-type locks have a number of pins (usually five) which must each be raised to a certain level by the key to allow the lock to operate; each pin usually has three possible levels, and a simple 5-pin key lock thus has 243 (= 3^5) differs; if the key's shaft also carries two long grooves that must match the lock's face plate and offer (say) a further 9 differs, the total number of differs is raised to 2187, which is considered adequate for all but the most stringent of security applications. In remote-controlled electronic locks, this level of security can be emulated with an 11-bit binary code or 7-bit trinary code.

Modern remote-control ICs specifically designed for general use in security applications such as electronic locks or burglar alarms have their code-frame bits divided into a number of 'address' bits and 'data' or 'address/data' bits. The address bits are used as an electronic key that can open a matching electronic lock; once the lock is opened, the data bits can control selected functions; address/data bits can be used as either address or data bits. *Figure 12.7* shows various ways of using the code bits of a 12-bit binary encoder/decoder system that has eight address bits and four address/data bits.

Figure 12.7(a) shows the basic 12-bit code. The 8-bit address code offers 256 possible key/lock codes; the remaining four bits of the code can be used as either address or data bits, in any desired combination. When the last four bits are used as data bits, they can be used to give four simultaneous control channels, or – with suitable decoding – up to sixteen non-simultaneous channels.

Figure 12.7(b) shows the 12-bit code used in an automobile remote-controlled central-locking and security-alarm-control application. Here, bits 1 to 11 of the 12-bit code are used as address bits and provide 2048 possible key/lock codes (for turning the solenoid-activated central-locking system on

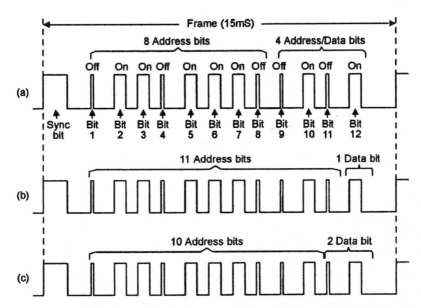

Figure 12.7 *Three alternative ways of using the outputs of a 12–bit binary coded encoder/decoder remote-control system in security applications (see text)*

or off), and the remaining data bit is used to turn the security alarm on or off when the central-locking system is on.

Finally, *Figure 12.7(c)* shows the 12-bit code used as a sensor's 'alarm' code transmitter in a wireless burglar alarm system. In this type of application, each sensor (a movement detector or a contact detector, etc.) can communicate with the system's main control unit via its own wireless transmitter, and is allocated one of four 'zone' or area-of-protection numbers. In the diagram, under an 'alarm' condition, the code's ten address bits provide 1024 possible identification codes which – if they match those stored in the main control unit – automatically activate the systems burglar alarm; the two data bits denote the sensor's zone number, which the main control unit automatically displays and stores under the alarm condition.

Code waveform types

The types of binary pulse-modulated waveform shown in *Figures 12.3* and *12.7* use a narrow pulse to denote a logic-0 (low or off) code and a wide pulse to denote a logic-1 (high or on) code; this waveform is thus known as a pulse-width-modulated (PWM) type. There are various other binary pulse waveform types. Among them are the pulse-amplitude-modulated (PAM) type, in which (at the moment in the code frame at which a pulse is scheduled to appear) a zero voltage amplitude denotes logic-0 and high amplitude denotes logic-1, and the pulse-position-modulated (PPM) type, in which logic-0 and logic-1 codes are represented by their relative positions in the frame.

One of the most useful code waveform variants is the 'dummy' type sometimes used in Holtek remote-control ICs. In these, the pulse frame may (for example) be an 18-bit type, using 8 programmable address bits and 4 variable address/data bits, plus 6 'dummy' bits that cannot be varied. This type of IC is as easy to use as a normal 12-bit type, but offers an apparent (to a criminal) 18-bit level of security (i.e. up to 262 144 apparent possible address codes).

Block diagrams

TV/Hi-Fi controllers

Figure 12.8 shows the typical block diagram of a multichannel IR remote-control transmitter of the type used with domestic TV and Hi-Fi units. The transmitter is fitted with a multi-function keyboard, which has its X and Y outputs repeatedly scanned via an encoder circuit that controls the input to a code waveform generator system. This latter unit generates the carrier wave signal (typically about 30kHz) and the 6–bit plus sync pulse repeating frame waveforms, which are then passed on to a standard IR transmitter output stage.

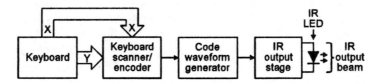

Figure 12.8   *Block diagram of a typical TV/Hi-Fi type IR remote-control transmitter*

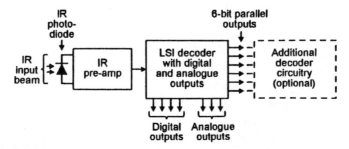

Figure 12.9   *Block diagram of a typical TV/Hi-Fi type IR remote-control receiver*

In the receiver circuit (*Figure 12.9*) the detected IR signal is first fed to a fairly sophisticated pre-amplifier stage which provides very high gain for long-range operation but does not saturate if the transmitter is used near the receiver. The pre-amp output is fed to an LSI (large scale integration) decoder IC, which typically directly provides three or four digital outputs (simple on/off functions) and two or three analogue outputs (volume, brilliance, etc.), but also provides a 6-bit output that is a parallel-coded version of the original 6-bit serial code and can optionally be decoded via additional ICs to give more remote-control functions.

'Security' controllers

Figure 12.10 shows the typical block diagram of a 12-bit IR remote-control transmitter (Tx) of the type used in many modern 'security' applications. In this example, the 12-bit code uses eight address bits (A0 to A7) which are permanently switch-wired into the unit via the S0 to S7 miniature switch bank, and four data bits (D0 to D3), which can be activated manually via switches S8 to S11 (in this example, the data terminals are internally biased high when their switches are open, and are pulled low when their switches are closed). The oscillator and code waveform generator circuit repeatedly samples the states of the 12 switches and converts the data into a sequence of 12-bit (plus sync bit) data frames that are transmitted in serial form via the 'Out' terminal; the data is converted into IR form via an internal or external IR output stage.

Figure 12.11 shows the typical block diagram of a 12-bit IR remote-control receiver that is designed for use with the above 12-bit Tx unit. Here, the IR signal is detected and converted into clean pulse form via the IR pre-amp and fed (in serial form) to the input of the comparator/decoder unit, which

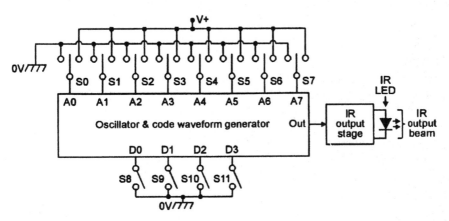

Figure 12.10 *Block diagram of a typical 12–bit 'security' type IR remote-control transmitter*

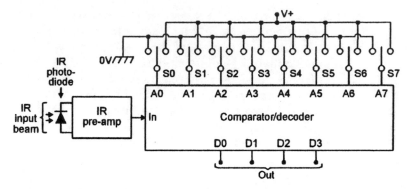

Figure 12.11   *Block diagram of a typical 12–bit 'security' type IR remote-control receiver*

converts each frame into parallel form and compares its 8-bit address with that set internally by the receiver's S0 to S7 miniature switch bank. If the two 8-bit address codes are the same, the receiver makes the four transmitted data bits (D0 to D3) available on the receiver's four output pins; if the two addresses are not the same, no outputs are made available (in some systems, the receiver goes into a temporary time-controlled shut-down mode under the latter condition).

Practical remote-control systems

TV/Hi-Fi controllers

During the 1970s and 1980s only fairly expensive TV and Hi-Fi units were supplied (as standard) with IC-based IR remote-control systems, and many electronics DIY enthusiasts thus took great interest in such ICs and systems at that time. Several ICs of this type were generally available in that era, the best known of them being the '490/922' 32–channel range of devices from Plessey, and the 'IR60' 60-channel range of ICs from Siemens, both of which were extensively described in the first (1988) edition of this book. Today, however, these particular ICs are no longer produced, and the demand for DIY TV/Hi-Fi remote-controllers has (for fairly obvious reasons) completely collapsed; consequently, no additional information is given on this subject in this edition of the book.

'Security' controllers

Today (1999), there is much DIY interest in IR security controller ICs and systems, and many such ICs are thus generally available in this highly competitive and innovative market. A major problem with all of these ICs, however, is that they have very short production lives (usually less than two years), and tend to become obsolescent or to have ceased production by the time they appear in dealer's catalogues. This factor makes it unwise, in a book of this kind (which has an intended production life of several years), to present 'dedicated' circuit designs based on such ICs.

Consequently, this section simply presents a few useful 'for guidance only' application circuits, plus a few tips on using such ICs, which use a minimum of external components and are generally very easy to use. Readers wishing to see application circuits and other data for current-production security controller ICs should consult manufacturer's current application manuals or visit their Web sites on the Internet.

Figure 12.12 shows the typical application circuit of an 18-pin 12-bit encoder IC that has eight address (A) bits and four data (D) bits and operates in the same basic way as the *Figure 12.10* system. In this case the address bits are set by the S0 to S7 miniature switch bank, and the data bits are set by manually operated switches S8 to S11. Resistor Rosc (typically 100k to 1M0) sets the IC's internal oscillator frequency. The IC's serial output appears on pin-17 and must be fed to an external IR output stage. The pin-14 LED (which has its current limited by resistor R) illuminates when a serial code is being transmitted.

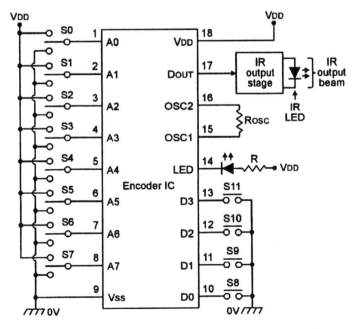

Figure 12.12 *Typical application diagram of a 12–bit 'security' type encoder IC*

Figure 12.13 shows the typical application circuit of an 18-pin 12-bit decoder IC that is designed to complement the above encoder system and operates in the same basic way as the *Figure 12.11* system. In this case the encoded IR beam is picked up and amplified by an external IR pre-amp and is fed into pin-14 of the IC, which has its internal oscillator frequency set by resistor Rosc (typically 100k to 1M0). The IC's eight address bits are set by the S0 to S7 miniature switch bank. When the address code of a received IR transmission matches that of the decoder IC for two or more successive frames, the IC presents the transmitted four data bits on its pin-10 to pin-13 outputs; simultaneously, the IC's pin-17 output switches high, to indicate the reception of a valid transmission (VT) code signal.

The two ICs described above are designed for use with any type of transmission medium (wired, wireless, IR, etc.); the connections shown apply specifically to a pair of now-redundant Holtek ICs (the HT6014 and HT6040), but are still quite typical of the present generation of ICs. All such ICs use a switch bank to set their address codes, have their internal oscillator frequency set via a couple of control pins and one or more external components, and receive or transmit their serial codes via a single pin. Some use trinary rather binary codes, some have active-high outputs and others have active-low outputs; most transmit their outputs in pulsed-tone form, but a few transmit them in plain pulse form.

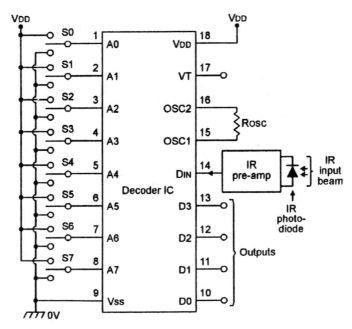

Figure 12.13 *Typical application diagram of a 12–bit 'security' type decoder IC*

When experimenting with ICs of these types, start by getting copies of their data/application sheets, and *carefully check the application diagrams for obvious errors* (such as nine address switches on an IC that uses only eight address bits, etc.). When initially checking the interaction between an encoder and decoder IC, start by powering them both from the same supply and using a wired link (plus a 10k series protection resistor) between the encoder output and the decoder input. When the ICs works correctly in this mode, you can progress to designing/building the system's IR data link.

IR data links

Transmitter links

Figure 12.14 shows two simple IR transmitter circuits that can each be driven by the pulsed-tone serial output of a security-type encoder IC and use the same power supply as the IC (minimum value = 4V). The *Figure 12.14(a)* circuit is designed for use with active-high pulsed-tone signals. Here, when the input is low, Q1 is biased off by R2, and no current flows in the IR LED, but when the input is pulsed high the pulse's tone signals repeatedly switch Q1 on and off for the duration of the pulse, thus feeding switching drive currents through the IR LED.

The *Figure 12.14(b)* circuit is designed for use with active-low pulsed-tone signals. Here, when the input is high (within 500mV of V+), Q1 is biased off by R2, and no current flow in the IR LED, but when the input is pulsed low the pulse's tone signals repeatedly switch Q1 on and off for the duration of the pulse, thus feeding switching drive currents through the IR LED.

In *Figure 12.14*, the peak operating current (Ipeak) of the IR LED (in amps) equals approximately (V+ − 3V)/R3. Thus, Ipeak = 0.2A when V+ =

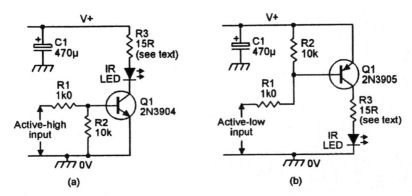

Figure 12.14 *Simple IR transmitters designed for use with pulsed-tone (a) active-high and (b) active-low inputs*

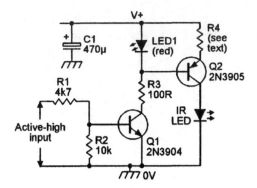

Figure 12.15   *Improved IR transmitter designed for use with an active-high pulsed-tone input*

6V and R3 = 15R. In practice, Ipeak should be limited to a maximum value of 0.2A by suitably increasing the R3 value if V+ values greater than 6V are used. Note that C1 acts as a low-impedance energy store and performs a vital operating function in these circuits, which can each use TIL38 or LD271 (or similar) IR LEDs (see *Figure 11.8*, in Chapter 11, for physical details of these IR LEDs).

A minor weakness of the *Figure 12.14* circuits is that their IR LED operating currents (and thus their IR beam intensity values) vary with the supply voltage value. *Figures 12.15* and *12.16* show improved versions of the circuits; these versions do not suffer from this defect.

The *Figure 12.15* circuit is designed for use with an active-high pulsed-tone input. Here, when the input is low, Q1–Q2 and the IR LED are off, but when the input is high Q1–LED1–Q2 and the IR LED are all driven on. Under this latter condition about 1.8V is developed across LED1, and 0.6V less than this (1.2V) is thus developed across R4; consequently, since the R4 voltage is determined by Q2's emitter current (which is almost the same as Q2's collector current), Q2 acts as a constant-current generator under this condition and feeds a stable peak current (Ipeak) of 1.2/R4 amps into the IR LED. Thus, Ipeak = 0.1A when R4 = 12R, or 0.2A (the maximum recommended value) when R4 = 6R0.

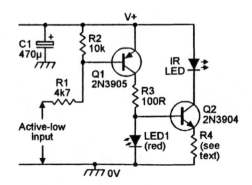

Figure 12.16   *Improved IR transmitter designed for use with an active-low pulsed-tone input*

Figure 12.16 shows an 'active-low pulsed-tone input' version of the above circuit; the circuit's operating theory is fairly self-evident from the above sets of descriptions. Note that these two circuits require an absolute minimum operating voltage of 4V; if voltages greater than 6V are available, two IR LEDs can be wired in series (see *Figure 11.9* in Chapter 11) to double the available IR output power.

In cases where the serial output of the security-type encoder IC takes a simple pulsed (rather than pulsed-tone) form, this output should be used to gate a simple CMOS-type 30–40kHz astable, and the astable output should then be used to drive one of the *Figure 12.14* to *12.16* circuits and thus generate an appropriate pulsed-tone IR output.

IR pre-amplifier circuits

In an IR data link, the transmitted IR signal must, at the receiver end of the system, be detected and amplified via a suitable pre-amplifier circuit before being fed to the input of the decoder IC. This pre-amp must, if a fairly long operating range is required, be reasonably frequency-selective (to respond specifically to the IR pulsed-tone signal and reject unwanted noise) and must provide very high signal gain. To conclude this volume, *Figure 12.17* shows a practical example of such a circuit; ideally, this circuit should be mounted within a screened case, to minimize the unwanted effects of electromagnetic interference.

The *Figure 12.17* design uses readily available components and is based on the *Figure 11.13* circuit described in Chapter 11, but used only a single IR photodiode (IRD1) and has some of its component values changed, to centre its frequency response on 30kHz. IRD1 can be any of the types shown in *Figure 11.14* (in Chapter 11).

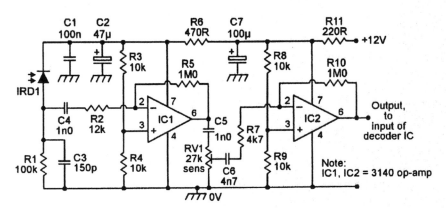

Figure 12.17 *30kHz IR pre-amp, using readily available components*

Index